Born of Ice and Fire

Born of Ice and Fire

How Glaciers and Volcanoes (with a Pinch of Salt)
Drove Animal Evolution

GRAHAM SHIELDS

Yale UNIVERSITY PRESS

New Haven and London

Published with assistance from the foundation established in
memory of James Wesley Cooper of the Class of 1865, Yale College.

Yale University Press books may be purchased in quantity for
educational, business, or promotional use. For information, please
e-mail sales.press@yale.edu (U.S. office) or sales@yaleup.co.uk
(U.K. office).

Set in Minion type by Integrated Publishing Solutions.
Printed in the United States of America.

ISBN 978-0-300-24259-1 (hardcover : alk. paper)
Library of Congress Control Number: 2023931531
A catalogue record for this book is available from the British
Library.

This paper meets the requirements of ANSI/NISO Z39.48-1992
(Permanence of Paper).

10 9 8 7 6 5 4 3 2 1

In awe of myriad unsung contributions to human knowledge

All life is parasitic;
like a swirling eddy it feeds off the energy cascade

Contents

Preface

Ever since Darwin, geologists have been trying to reconstruct the evolution of life from our woefully incomplete fossil record. One of the most vexing problems has been the apparent absence of any long fuse to the explosion of animal forms about half a billion years ago, during the Cambrian Period. Fossil evidence for life's gradual march toward biological complexity has been a long time coming, but recent discoveries have come thick and fast. Now that several key missing pieces of the puzzle have finally been unearthed, it is perhaps time to take stock of what we know about how and why our earliest animal ancestors first emerged.

This book was naively spawned at a time when neither I nor any of my fellow scientists knew when the first animal evolved, into what conditions it emerged, or even what it looked like. Now, many years on, I think I know at least some of the answers well enough to feel more confident putting pen to paper. In what follows I have tried to sketch out how and when conditions enabled animals to evolve, radiate, and diversify into the myriad forms that exist today. I have also tried to reach beyond our specific origins to draw wider conclusions about the probability of complex, energetic life forms on a habitable planet.

Although the origin and evolution of animals is very much
the focus of this book, it is a book not of bioscience, but of
geoscience. I leave it to others to tease apart the order and significance of specific animal groups, their genetic machinery
and defining characteristics. I am interested here not so much
in the precise means by which some single-celled protist more
than 600 million years ago made the crucially important steps
toward complex multicellularity, but in how the environmental
landscape changed to make such a new way of life profitable.
In recent years, our understanding of the physical evolution
of the earth's planetary surface has greatly improved, shedding
light not only on the dynamic interactions between life and the
environment but also on the driving forces behind evolutionary progress.

This is a book of science from the perspective of an academic who has picked at the coal face of this subject for most
of a career. It has been a labor of love and a salutary lesson in
how scientific knowledge grows. The geological slant may be
new to many readers who, like me, will be people with curious
minds, eager to add nuance to received notions. As such it is
different from other books on early life, as my perspective falls
squarely on a view of the earth as a complex, emergent system,
comprising the biosphere, geosphere, atmosphere, and hydro-
sphere, which needed to evolve, just like life, into its modern
form. Today, our world is characterized by its equable climate,
oxygenated oceans and atmosphere, and advanced, energetic
life forms, but this was not always the case. This book narrates
a journey to modernity that navigated wild swings between hot
and cold climates, oxygen-replete and oxygen-starved conditions, biological radiations and extinctions, and asks how such
tremendous instability might relate to even grander tectonic
cycles.

I am an avid reader of science, and I hope that this book

will be of interest to other curious minds who wish to explore how the earth's present (and future) is dictated by its past. It is a subjective account peppered with my own anecdotes and experiences, but one that I trust serves up a balanced menu of personal interpretations and latest findings from a diverse range of fields from sedimentology to tectonics and from geochemistry to paleontology. It also contains historical accounts of various heroes of mine: scientists who dedicated their largely unsung lives to enlarging the edifice of modern science. Many of the ideas in the book are my own, some published in peer-reviewed journals and some yet to be so scrutinized. In that regard, I would like to think that my scientific peers may also find my musings of academic interest.

The question of how we humans and our direct line of animal ancestors came to evolve on this planet was until recently a philosophical more than a scientific question. But so much has happened recently to change all that and to make this book possible. A combination of fossil and molecular evidence has traced our origins to an intriguing chapter in the earth's history when the entire globe was buried in thick ice for tens of millions of years. This "Cryogenian Period" (from the Greek for "cold" and "birth") also bears the evocative name "Snowball Earth." The time before, during, and just after that frigid episode witnessed extreme changes not only in climate, but also in tectonics, oxygen, nutrients, and evolution that we are only now starting to understand. If nothing else, this book tries to get to the bottom of what caused that uniquely disruptive time in earth history, and how such instability contributed to our own origins.

Scientists are often treated with bemusement or even suspicion when they cautiously express their ideas in probabilistic language. I have also noticed how uncertainty can be scary for my students, who often begin their studies simply wanting to know the answers they need to pass their exams. However, the capacity to change one's mind is an essential part of what it means to be a scientist. Science is, after all, less a roll call of facts and more the measurement of doubt. Because we should view all uncertain conclusions with a healthy pinch of salt, I have tried to set out primary, largely factual geological observations at the outset, gradually adding the weight of circumstantial evidence from complementary disciplines, to reach a somewhat lesser degree of uncertainty by the close. Although so much still remains to be discovered about how animals came to evolve, I am encouraged that we have come so far in such a short time.

Acknowledgments

In writing this book, I am greatly indebted to a number of friends and colleagues without whom it would be very much the poorer. My wife and fellow scientist Ying Zhou has supported me through the entirety of this journey, intellectually and emotionally, ever confident that I will eventually reach the light at the end of this surprisingly long tunnel. For numerous stimulating discussions I am greatly indebted also to various scientists who have had to put up with my obsessions and entertained me with some of their own. In this regard, I'd mention in particular Simon Poulton, Rachel Wood, Tim Lenton, Fred Bowyer, Richard Boyle, and Maoyan Zhu. A lot of the ideas toward the end of this book derived from brainstorming sessions outside my office or in various watering holes with Ben Mills, who added much needed legitimacy to my arithmetic scribble. Nick Lane has also been a constant source of inspiration and ideas, ever willing to grapple with my geochemical ponderings, usually over lunch and a beer. It is too difficult to name all of the geologists who have inspired me. Their legacy can be found in numerous books and memoirs that form the foundations of any geological study. Some of them are named here in the book, and I'd single out a few of those as personal mentors whose attention to detail and veracity I still hold dear.

In the approximate order in which we met, they are: Ian Fairchild, Ken Hsu, Martin Brasier, Max Deynoux, and Tony Spencer. Specific thanks also to Ying Zhou, Rachel Wood, Elias Rugen, and an anonymous reader for reading through earlier drafts of this book. I am grateful for their suggestions, which have significantly improved the contents. My final thanks go to the staff at Yale University Press for their unfailing professionalism and support. Especial thanks go to Joseph Calamia, who launched me on this journey, and to Jean Black, Elizabeth Sylvia, Erica Hanson, and Phillip King, who have seen it through to the end.

1

Time Travel

Shortly after I began my first teaching job, I remember standing self-consciously in a lecture room at James Cook University in Townsville, Australia, my arms outstretched as if ready to take off over the heads of about fifty bleary-eyed teenagers. It seemed to take a geological age before I was finally able to tear their attention away from their phone screens. Not wishing to lose the moment, I explained with missionary zeal that the earth's history covered unimaginably huge spans of time. So huge in fact as to render meaningless such phrases as "unimaginably huge." After all, I ventured, is a million years any more imaginable than a billion years? If we were to rise above the constraints imposed by our limited life spans, I told the students, then we needed a *metaphor*!

If you do the birdman impression yourself, try to imagine that your full wingspan comprises the entire history of our planet condensed into a human tickertape that stretches from one outstretched fingertip to the other. At one end, the earth came into being, thrown together by a series of violent collisions with myriad jostling neighbors all in a spinning disc encircling the embryonic sun. By the time we reach the wrist, the primordial earth has already been through hell and back, having survived impacts with other cosmic bodies, themselves as large as Mars. One such collision formed the moon. We call this handspan of the earth's history the *Hadean* Eon because if life ever existed this far back it would have been sterilized to oblivion after each devastating smash. Your other wrist is also of great significance, for it was then that the first animals finally managed to gate-crash the party. Your arms and chest represent, therefore, more than three billion years of uninterrupted dominion by microbes, and later algae. Flightless dinosaurs, everyone's favorite fossils, evolved around your knuckle, only to disappear from the face of the earth when a meteorite put an end to them at the last of your three finger joints. A mere fingertip is all that remains for those things that seem to have been around forever, like whales, primates, grass, the Himalayas, and the Rolling Stones. If you were feeling particularly despotic, you could wipe out the entire human race simply by filing your nails.[1]

You can put your arms down now. Phew. As metaphors go, this one is not bad, but it breaks down when you get to more recent times. I mean, who can envisage the time since the construction of the seemingly ageless pyramids as a mere 1% of 1% of a nail clipping? There are plenty of other metaphors that might be made to work even better. For example, if the history of the earth could fit into one whole year, then Charles Darwin published his theory of evolution by means of natural selection on the eleventh peal of the church bells at midnight on the last day of the year.

The tale in this book is about the period represented by your wrist, when the first signs of animal life appeared on the scene. I think it is the most extraordinary period in all earth history, and I invite you to accompany me on a journey back through time to our ancestors' most formative years. Despite their importance, until recently we knew relatively little about those times, and they are clearly less cinematic than other, more recent periods. Time travelers in fiction, for example, always seem to end up in the age of the dinosaurs. I suppose it fits with most people's idea of what the world was like in the dark and deepest past. In films, mad scientists step intrepidly out of their time machines only to be startled into panic by the roar of a ravenous *T. rex*. Isn't it always the way? The story of *Jurassic Park* paints a similar picture, but in reverse, with ferociously intelligent velociraptors taking to today's African jungle like fish to water. We take such fluidity for granted in works of fiction, and perhaps we should, because the worlds of us-and-them, although separated by tens of millions of years, are not so very different. After all, they lived only one finger joint away. If we could really go back in time, and I wish we could, I would gladly volunteer to be the first to breathe the agreeable fresh air of the Mesozoic Era.

However, fictional time travel can be misleading, as the world was not always so friendly to our metabolism. Our planet has existed for seventy times longer than the time span separating us from those voracious carnivores. For the first four billion years, informally called the Precambrian (or pre-Cambrian), but more formally comprising the Hadean, Archean, and Proterozoic Eons, each divided into periods, some yet to be named or defined, few organisms could even move, let alone run. The Hitchhiker's Guide to the early earth would probably not call it "mostly harmless," in the words of the late, great Douglas Adams. If the lack of oxygen did not kill you first, an evil concoction of poisonous gases or temperature extremes would soon see to your demise. Seafaring animals like fish would not fare any better. If the oceans contained any oxygen at all, it would have been only in the shallowest of realms. Slowly decomposing organic garbage would have turned the rest of the oceans into a foul-smelling, toxic stew. Bacteria, including many anaerobic types that thrive today only in hiding beneath the seafloor or inside your gut, were the main custodians of the planet for eons, until something caused all this to change (Figure 1).[2]

For some reason, after a billion years of relative stasis, dubbed the "boring billion," a biological revolution took hold. Some revolutions revert quickly to the status quo, like the French Revolution, which led to the weary quip *plus ça change, plus c'est la même chose* (the more things change, the more they stay the same). By contrast, other revolutions, like the Industrial Revolution, break the mold. They not only alter the course of history, but rewrite the rulebook as well. Because it is nigh impossible to uninvent an invention once it takes hold, new technologies reweave the fabric of society. Once fire, wheels, clocks, engines, and then computers entered our collective consciousness, humanity could never return full circle. The earth's great revolution, from a simpler microbial to a more complex biological world, was akin to this second kind of revolution. Our planetary home, once thrust into modernity, never looked back. New life forms arrived with new behaviors that changed the earth system in ways that we are only now beginning to understand.

Until recently, we could only speculate about what caused life's great "industrial" revolution. Indeed, ever since Darwin there has been no shortage of ideas to explain first the emergence and then the diversification of animals between about 600 and 500 million years ago. Some have preferred an intrinsic explanation, arguing that the requisite genetic machinery simply needed eons to evolve the necessary mutations. However, most have accepted that "permissive" environmental changes played their role, too. Many have argued—and I wholeheartedly agree—that it can hardly be a coincidence that the oceans became fully oxygenated for the first time before the Cambrian

Age (Ma)	Eon	Era	Period		Meaning	Significance
0	Phanerozoic		Cenozoic	visible life	*new life*	Age of mammals, grass; cooling
200			Mesozoic		*middle life*	Age of reptiles; calci-plankton
400			Paleozoic		*old life*	Age of amphibians, plants; ice age
						Age of arthropods, biodiversification
600	Proterozoic	Neo-proterozoic	Ediacaran	new early life	*Ediacara*	Ediacarian fauna; first animals; NOE
			Cryogenian		*icy cold*	Glacial deposits; iron formations
800			Tonian		*stretch*	Rifting; volcanism; protists emerge
1000			?Kleisian		*closure*	Supercontinent assembly: Rodinia
		Meso-proterozoic	Stenian	middle early life	*narrow*	Orogenic belts; metamorphism; green and red algae emerge
1200						
1400			Ectasian		*extension*	Continuing platform sedimentation
			Calymmian		*cover*	Tectonic lull; extensive platforms
1600			Statherian		*stable, firm*	First supercontinent (Nuna); first eukaryotic micro- and macrofossils
1800		Paleo-proterozoic	Orosirian	old early life	*mountains*	Widespread mountain building; metamorphism; first macrofossils
2000						
2200			Rhyacian		*lava flows*	Lomagundi C-isotope excursion; first giant sulfate deposits; widespread volcanism
2400			?Skourian		*rust*	Atmospheric oxygen accumulates; GOE; major glaciations
2600	Archean	Neo-archean	Siderian	new ancient	*iron*	Banded-iron formation acme; megacontinent assembly
2800			?Kratian		*strength*	Sedimentary basins developed on stable or growing cratons; first *whiffs* of oxygen
3000		Meso-archean		middle ancient		Growth of cratons with felsic / granitic upper continental crust; first microfossils
3200						
3400						
3600		Paleo-archean		old ancient		Earliest preserved sedimentary rocks and isotopic traces of life
3800						Oldest vestiges of crust
4000	Hadean			barren, unseen		Early crust formation & recycling; heavy meteorite bombardment
4200						
4400						Accretion of Earth; then giant Moon-forming impact event
4567						

Figure 1. The geological time scale. Geologists divide time into eons, eras, and periods, based on events recorded in rocks. Here you can see a recent attempt to do this for all of the earth's history (the numbers at left are the age, in millions of years ago). Period names are based on Greek words; those in italics are suggestions only and have not been ratified by the International Union of Geological Sciences.

"explosion." Notwithstanding the importance of oxygen for most if not all animals, there is, however, understandable disagreement over whether oxygenation was a cause or an effect of animal evolution. Unprecedented tectonic upheavals, involving the breaking apart and collision of mighty continents, formed the dynamic backdrop to biological innovation, suggesting perhaps that both life and environment were playing to the tune of something far grander in scale.

I was only a student when the germ of a seemingly crazy idea began to take root. It centered around controversial evidence that just before animals emerged, our planet had been so deeply gripped by glaciation that ice caps even reached the equator. By the end of the 1990s, this fancy had morphed into the popular "Snowball Earth" hypothesis, which caught the imagination of scientists and non-scientists alike. I feel awestruck at how much more we know now than we did in those heady days only two decades ago. Nowadays, this once ridiculed idea is so widely accepted that only the details are worth squabbling over. In my experience, nothing reveals better how knowledge grows than the Snowball Earth debate, which has seen its predictions forensically examined one by one by scientists across the globe. Intriguing though it is to imagine an ice-covered planet, what can such seemingly inhospitable conditions have to do with our own origins, if anything?

If it were not for recent fossil discoveries and genetic research, this question would remain moot. However, phylogenomics, the study of genetic differences among modern organisms, has pinpointed the Cryogenian deep freeze as the crucible from which all animal life subsequently evolved. Simultaneously, our knowledge of the fossil record has improved to the extent that we can now be sure that no modern animal groups were present before Snowball Earth, but plenty of evidence points to their emergence and radiation once the snowball had melted. Most recently, organic molecules, miraculously preserved across the eons, have joined the fray to highlight how this great ice age also witnessed the rise of complex, multi-cellular algae around the globe. The Snowball Earth chapter in our planet's history divides biological evolution into two almost entirely separate parts. Why these revolutionary events occurred, and how they can help us understand our own evolutionary beginnings, are key questions that will accompany us throughout this book.

The unbroken habitability of our planet links intimately with the evolution of the rest of the earth system, namely the atmosphere, oceans, and rocky interior, while all these together affect and are in turn affected by climate. More than ever before, humans are aware that we live in and form part of a huge and complex system, with each part connected to all others by material flux on a planetary scale. We know that the constant recycling of life's chief ingredients—carbon, oxygen, and the nutrient elements phosphorus and nitrogen—sustains organic productivity, governs atmospheric composition, regulates surface temperature, and controls many more things we simply take for granted. It is perhaps because of this that we find it hard to imagine a world before this one, as unrecognizable to us as is Mars. It is inconceivable today that our planet could ever descend into such perpetual deep freeze, yet those conditions must have been conducive, even essential, to our own origins. The fact that this did happen, albeit very long ago, means not only that such extreme events are possible, but that they were a natural condition for our planet. Earth languished beneath its frosty blanket for an immensity of time, far longer even than the interval separating us from the flightless dinosaurs. To understand such a world is to recalibrate our idea of what "normal" is, and perhaps to abandon the uniformitarian mantra of geology that "the present is the key to the past."

The great strides in understanding that have been made over the past twenty years illustrate well how the Precambrian earth challenges our thinking because they simply don't fit with what we know from the modern world. From extreme climates to inexplicable changes to ocean chemistry, little makes sense when viewed backward from a modern earth system perspective. How can ice have reached the equator anyway without triggering negative feedbacks that we know from the modern world should set global climate back to rights? Surpassing our failures of imagination requires us to adopt a different way of thinking: one that integrates all of these apparently disconnected phenomena into a narrative that makes sense when we let geological clocks run forward. A historian would not try to understand the Industrial Revolution by analyzing computers, and it is important not to make the same mistake in geology by using only today's world to understand the earth's deep history.

Geology is the ultimate cold case, wherein geological detectives forensically piece together the evidence from only the few tiny jigsaw pieces remaining after the ravages of time. As a result, this is a book of two distinct parts. In the first part I outline the details and quality of the evidence that has been teased out of the imperfect rock record. As in any whodunnit, the best way to establish cause-and-effect is to make a timeline of events, but this is easier said than done when the events happened more than half a billion years ago. Nevertheless, if we are to make any sense at all of the huge amounts of data that are continually being amassed, putting our knowledge in time order is absolutely essential. Until recently, however, such a task was nigh impossible.

Since the 1990s, geology has become increasingly collaborative and multidisciplinary, with new data arising all the time from chemistry, physics, and biology. As a result, we can now attempt—as I do in the second part of this book—a plausible, defensible, and testable synthesis for what caused animals to emerge when they did. Details will undoubtedly change, but I trust that the essence of my conclusions will stand the test of time: that our earliest animal ancestors were spawned from ice, cradled by fire, and fueled by the weathering of rocks, one of the most important of which turns out to be gypsum, the soft, white, powdery mineral from which alabaster, plaster, and blackboard chalk are made. Here again, we must abandon our precious uniformitarianism, this time with regard to the workings of the modern carbon cycle. If we are to understand the carbon (and oxygen) budgets before the rise of animals, we need, counterintuitively, to look to sulfur. Behind all these part-time actors in the narrative of our planet stand grand phenomena of paramount importance, for without the sun and tectonics, there would have been no revolution at all, and no one to tell the tale.

Our existence on this planet is contingent on many different past events, but if we were to fire time's arrow once more, would we humans still exist? Probably not, for there are just too many pivotal coincidences that would need to swing in our favor. Would intelligent life of any kind exist, or would animals have evolved at all? Would there even be sufficient oxygen to drive their energy-sapping metabolisms? These are more interesting questions because their answers depend not on an unbroken series of improbable events, but instead on whether the improbable is inevitable given enormous spans of time and certain starting conditions. I sense that evolution's trajectory is contingent not only on catastrophe and opportunity, but also on long-term planetary-scale forces that over geological eons will eventually and inevitably prevail. Which of these drivers, if any, was truly essential for the evolution of intelligent life? The answer to that question lies far back in geological time and may well surprise you.

Before we delve even deeper into the earth's ancient past, let's begin our journey by visiting the icy cradle of our own animal origins. In other words, let's first travel back only as far as the wrist before investigating what happened before or after life's great revolution. If the present is not always the key to the past, perhaps it makes perverse sense that I am taking you first to one of the hottest places on earth: the Sahara Desert.

2

Saharan Glaciers

For a century or more, evidence has mounted showing that ice caps were widespread shortly before animal life forms radiated during the Cambrian Period. Although detailed surveys by unsung geologists around the world have put the case for glaciation beyond dispute, ripples of skepticism have followed every claim that the Precambrian glaciation was unusually extensive. Ironically, a good place to start resolving this icy quandary is the Sahara Desert.

"Whatever you do," Max warned, "don't take the Nouakchott taxi!" But now, it was too late to change my mind as I watched the man, who had gleefully taken my money, with mounting incredulity. Over the past two hours I had seen him guide the loading of wares, including an entire car engine, onto the now sagging roof of the decrepit Land Cruiser. It was only when I finally got on board that I realized that this old gent, with his withered arm and glass eye, would also be our driver.

Abruptly, we left the capital far behind us, and the seemingly endless landscape of bare sand and rock gave me plenty of time to marvel at how all of us had managed to squeeze inside. Clearly, thirteen was not an unlucky number in Mauritania. Three women with flowing blue robes were sprawled, almost elegantly, alongside but somehow not quite touching the driver, while I was in the back seat squashed tightly against a gaunt, elderly man who howled in pain each time his head bounced off the ceiling. Just when I thought it could not get any worse, he negotiated a seat swap with a burly mechanic from Dakar. Eleven grueling hours later, with aching bones and fraying tempers, we finally reached the desert city of Atar.

Being by necessity an outdoors and increasingly global pursuit, geology has a tendency to both dismay and delight. I remember the time in Senegal when a troop of baboons scared me half to death by racing toward me from the valley's far side. From their new vantage point, they could now throw rocks at my head while I scanned the cliff face with the tools of my trade, a trusty hammer and a hand lens. Or the time I went swimming in northern Australia, to wash off the day's grime, only to discover that the floating logs were in fact crocodiles ("No worries, only freshies," quipped my more seasoned Aussie colleagues). It seems that the geology I pursue, of which we will hear more in due course, is often found in far-flung, exotic places, like the Australian outback, the Yangtze Gorges, and the Mongolian taiga. Even the places I haven't been, but where rocks of the same age can be found, like Namibia, Siberia, Yukon, Death Valley, and northern Greenland, are hardly your average holiday destinations. Writing in the comparative comfort of my snug London office, the memory of that hot and sweaty day in the Sahara Desert still makes me smile and grimace in equal measure.

The story I want to relate in this book concerns all of these wonderful places and just some of the people who have dedicated their lives and careers to their study. The science of geology is replete with unsung heroes: men and women who have spent countless years treading foreign parts in order to conjure up a three-dimensional picture of the rocks beneath our feet. In so doing, every geologist has in some way, large or small, helped to reconstruct by increments the fascinating story of how the modern world came into being. The man I wanted to meet up with in the middle of the Sahara was one such person. By the time of my visit in 1996, Max Deynoux had spent more than thirty years mapping out the Sahara Desert in painstaking detail. When I finally arrived in Atar, a small town of about 25,000 inhabitants and the capital of the Adrar region of Mauritania, Max was waiting for me at our agreed-upon rendezvous point. A few months earlier, he had told me that he had almost given up on finding a geochemist who would venture out with him into the field (many don't seem to leave their labs except to go to conferences), and I didn't take much persuading. I wanted to learn from Max firsthand what he knew about one of the most intriguing turning points in earth history. I was not disappointed.

Max, a keen rugby player in his youth, was a fit, strong, and exceedingly proud Frenchman. Two decades earlier, in 1976, he had gone from relative obscurity to modest notoriety, at least among geologists, as the author of an apparently unremarkable scientific paper in the *American Journal of Science*. Co-authored with Roland Trompette, the paper is unremarkable in the sense that it is merely an English translation of a "comment" made about another paper published two years earlier. Normally, such things are destined to be forgotten footnotes in the annals of science. But Max's discoveries were to bring to light in unprecedented detail an episode of earth history that had hitherto remained as controversial as it was puzzling. His findings would help to unveil a picture of the ancient earth, draped in ice from pole to pole, where glaciers could touch the sea even at the equator. This period in earth history would come to be referred to as "Snowball Earth," but all that was yet to come.[1]

Although for me 1976 marks the turning point, evidence for an unusually severe ice age had been mounting for many years, with the first hints as long ago as 1871. In that year, James Thomson spoke to the forty-first meeting of the British Association for the Advancement of Science in Edinburgh on the subject of the geology of Islay, an island off the western coast of Scotland famous for its whiskies. In typical Victorian prose, Thomson concluded his address with the hope not to "overstep the bounds of prudent speculation in suggesting that [some boulders he had found there were] the reassorted materials of some great Northern Continent that has yielded to the ceaseless gnawing tooth of time . . . and been transported by the agency of ice." Thomson's evocative speculation about a "great northern continent" was well founded, it turns out. We now know that these red granite boulders came from North America, which back then was much closer to Scotland than it is today. As for the ice, Thomson was right there, too.[2]

By the 1960s, evidence that parts of our planet had been glaciated some time before the Cambrian Period—the time in earth history when animal fossils first become visibly abundant— had been reported by geological teams all around the world. Following those initial reports from Scotland others emerged, first out of Norway, and then Australia, Greenland, Svalbard, Namibia, the United States, China, India, Russia, and northwest Africa, all before the outbreak of World War II. One of the earliest proponents of a widespread Precambrian glaciation was the adventurer Douglas Mawson. Although Mawson is best known as one of the great Antarctic explorers, it is perhaps less

well known that he then became a professor of geology and
mineralogy at the University of Adelaide in Australia. In 1949
he proposed a global "late Precambrian ice age," largely on the
basis of what he had seen of Australia's geology. Unfortunately
for Mawson, not long after he formulated those ideas, Alfred
Wegener's once ridiculed theory of wandering continents gained
mainstream acceptance. The new postwar generation of geologists agreed that Australia had drifted to its present location
from much higher latitudes, having once been joined to Antarctica. Far from being anathema, ice caps in Australia were
fast becoming the expected norm. Against this already hostile
background, the notion of a global ice age during the late Precambrian would then be challenged from an entirely different
perspective.

Ever since the middle of the nineteenth century, the lay-
ing of telegraph cables across the Atlantic Ocean had facilitated
communications, but on November 18, 1929, disaster struck when
twelve cables broke, all on the same day. It was not until 1952
that two geologists from Columbia University, Bruce Heezen
and Maurice Ewing, pieced the mystery together. In a now classic paper, they reconsidered the evidence and noted that all of
the broken cables had been downslope of Newfoundland, where
earlier that same day a powerful magnitude 7.2 earthquake had
struck. Their conclusion was startling and would greatly influence how geologists view sedimentary rocks. In short, they
surmised that the earthquake had generated a submarine landslip, causing a devastating turbidity current of mud and rock
fragments that hugged the seafloor, traveling at speeds of up
to forty miles per hour. They estimated that as a result of just
this one shelf collapse in 1929, more than 200 cubic kilometers
of debris traveled over 1,100 kilometers, breaking the cables in
perfect time sequence from north to south. The cables did not
stand a chance. This was the final proof that turbidity currents

exist and caused a second headache for Mawson's late Precambrian glaciation because it had been assumed that the occurrence of large "erratic" boulders on a relatively flat, otherwise muddy seafloor was a uniquely glacial phenomenon.[3]

In the face of mounting skepticism, one of the earliest converts to continental drift attempted a counterinsurgency by establishing firm diagnostic criteria to distinguish ice-transported debris from current-transported debris. Brian Harland was a professor of geology at the University of Cambridge during the heady days of the plate tectonics revolution of the 1960s. Buoyed by newly developed geophysical techniques, he reiterated Mawson's conclusion of a late Precambrian glaciation, naming it the "Great Infra-Cambrian Glaciation" in 1964. We will revisit Harland's pioneering geophysical work later, but in short Harland's cleverly assembled arguments and new data did nothing to silence the cynics. To some degree, his own objectivity was to blame for the lack of impact his findings had. Harland divided glacial criteria into three categories: (i) those that are merely suggestive of glaciation; (ii) those that are decisive if found in sufficient degree or association; and (iii) those that are decisive on their own. The only feature that he dared place into the very important third category was evidence for ice rafting in the form of "dropstones," which are lone pebbles that have apparently fallen into muddy sediment from a passing iceberg. Geologists recognize them by the way they pierce or disrupt sedimentary laminae. Although icebergs certainly suggest glaciation somewhere else, even unambiguous dropstones did little to bolster the evidence for global glaciation.[4]

After a string of publications, Brian Harland and others before him had convinced many that there had been a late Precambrian glaciation of sorts, although perhaps of only patchy extent, so that by the 1970s, opinions were still very much divided. With the new paradigm of plate tectonics, geology had come into its own as a discipline, and many considered that the evidence for the Great Infra-Cambrian Glaciation was exaggerated or just plain wrong. One such skeptic was a geologist, then living in Portugal, named Lodewijk J. G. Schermerhorn. By 1974, Schermerhorn had amassed enough evidence to publish a withering 152-page critique that attacked the Mawson-Harland late Precambrian glaciation head-on. By forensically examining every known deposit worldwide, Schermerhorn concluded that glaciation had been limited to mountainous regions only, if it had existed at all. Schermerhorn's skepticism had the opposite effect of the one he intended. His paper sparked a string of papers and even books that finally released the cavalcade of evidence for globe-spanning glaciation that had been painstakingly gathered over the past decades, and kept buried within survey reports, theses, and obscure journals all over the world. The first and arguably most decisive nail was hammered by Max Deynoux.[5]

On the morning after the bruising journey from Nouakchott, Max was up bright and early, and curiously impressed albeit somewhat puzzled that I had entirely ignored his advice not to take the taxi. The previous evening, I had arrived in the dark, and apart from a hasty bonsoir to the hostel gatekeeper I had seen nothing and no one since arriving. I needn't have worried. Max took me straight to the market where I was immediately assaulted by the sights, sounds, and smells of a full-blown Saharan bazaar. After shopping for food—to be cooked each night by a local guide—we set off into the desert. My second night in the Adrar would be a very different affair. Eating steak, drinking wine, and constantly removing objects out of the path of wandering sand devils, we talked into the night about the nomadic desert tribes and the wild dogs we could hear howling around us. A rather too copious quantity of red wine had me attempting to translate Monty Python sketches into ropey

French, to the bemusement of Max and his wife, Janine, and to my eternal embarrassment the next morning.

There are some outcrops so rewarding that they become sites of pilgrimage for geologists around the world. Jbéliat is one such place, which gloriously rejects Schermerhorn's glacial reservations without need for words. To paraphrase Deynoux and Trompette's fourteen-page comment: Dr. Schermerhorn, if you still don't accept the evidence for late Precambrian glaciation, then you clearly haven't been to Jbéliat.

Sedimentary rocks are almost always found as horizontal layers of variable thickness called beds that record the times when sediment was deposited on the seafloor. Beds can represent events, such as the passing of a turbidity current. The diminishing strength of the current with time and distance allows ever lighter debris to be deposited. Any subsequent event leads to a new bed being deposited, and a layer cake (or stratigraphy) builds up, with each event bed marked by normal grading—that is, repeated cycles of upward-fining grains. Closer to the source of an event, however, the grading can be less obvious, or not even present, and large boulders and smaller cobbles and pebbles float higgledy-piggledy within a muddy matrix. Such debris flows leave diamictite deposits (massive fine-grained rock with outsized pebbles) that can be very similar to glacial till; hence the battle between Schermerhorn and Deynoux. Nevertheless, they can still be told apart if you know what to look for, and Max was keen to show me.

At Jbéliat, there is a horizontal surface that occasionally forms the flat desert floor we raced along in our gleaming white Land Cruiser. With only the four of us inside, I could finally enjoy the view and noticed that the surface also showed up as a perfectly straight line in the cliff face of the Falaise d'Atar. Away from the cliffs, the planar surface stretches for hundreds of kilometers into the desert and is an ancient landscape that you could easily have walked across 635 million years ago, just before it was inundated by the sea. In many places, the surface is overlain by a diamictite, just a few meters in thickness, comprising mud and boulders up to a meter or more across. You can follow this deposit for over a thousand kilometers to the northeast with barely a change in thickness. The exceptional thinness and singular nature of this deposit as well as its huge areal extent over an exceptionally flat peneplane rule out any similarity to debris flows, which tend to be much thicker, less regular, less extensive, and normally consisting of multiple events rather than just one. Many of the transported pebbles came from the north and had clearly traveled huge distances to get where they are today. All this points to a glacial origin . . . with a bit of imagination . . . but at Jbéliat there is no need to rely on imagination. One need only look down.

The surface of the desert at Jbéliat is more spectacularly and unambiguously glacial than any I have ever seen, as a photograph taken by Max in the early 1970s shows (Figure 2). It is highly polished in places, but also criss-crossed with scratches and grooves that generally reveal a north–south orientation. During many years of research, Max was able to trace this ancient surface over thousands of square kilometers. The surface is deeply scored by crescent-shaped grooves where the ice has plucked away the underlying rock basement. In some places, the diamictite is also grooved in the same direction as the underlying pavement, but less distinctly, showing that the passage of ice has left occasional hills of moraine in its path, just like in glaciated regions all over the world today. In still other places, where the underlying bedrock is harder, the basement stands proud as a mound of polished stone, asymmetrical in cross-section and resembling the back of a whale. These

Figure 2. Glacially striated pavement, Hank, Algeria, in
the Sahara Desert. (Photo by Max Deynoux, early 1970s)

hummocky features are uniquely glacial in origin and were
named *roches moutonnées* in 1786 by Horace-Bénédict de Saus-
sure, the Genevan geologist and founder of Alpinism.

Mouton is the French word for sheep. Whether de Saus-
sure was recalling a field of sheep or, as some would have it, the
wigs, smeared with mutton fat, that were popular in his day is
not entirely clear. It might even be that he was amusingly referring to both at the same time. What we do know is that their
characteristic whaleback form is a typical feature of landscapes
sculpted by ice and is due to abrasive polishing on the shallow,
smooth (stoss) side, which is always grooved in the direction
of ice movement, while the steep (lee) side is far rougher, having been eroded by the plucking action of passing glaciers. Max
showed me how to run my fingers along the polished grooves
to tell in which direction the glaciers had traveled. The mut-
toned wigs of the Sahara Desert are such classic examples that
they could have come straight out of a modern textbook on

Alpine glaciology. They confirm beyond doubt that the ice, like the boulders, came from the north.

Overlying the vast, striated pavement, the diamictite can be identified as a lithified moraine or tillite by the shape and texture of the large, outsized pebbles, or clasts, that sit in its friable, muddy matrix. The clasts are of highly variable composition and exotic provenance, having come from different localities both locally and up to several hundred kilometers away. Many of them were plucked from a glacial pavement, like the one beneath our feet, as they display sets of scratches and grooves, occasionally crossing each other, but typically parallel to the long axis of the clasts, which resemble many-sided flat irons with blunted edges. Most intriguingly, the clasts, and only the clasts, are coated in a thin, white skin composed of the mineral calcite, a form of calcium carbonate, something which Max informed me was also a typical feature of glacial moraines today.

Another well-known feature of modern, frigid terrain is patterned ground. Flying over Arctic regions, the polygonal patterning of the landscape made by the freezing and thawing of permafrost is plain to see. It seems almost inconceivable that such patterning could be preserved in the rock record, but Max was convinced that he had found some. At the top of the diamictite, Max had discovered an extensive layer of dark sand, which from the air can be seen to form polygonal structures of some seven to ten meters in diameter. The polygonal shapes demarcate sand-filled, wedge-shaped cracks that clearly split the underlying, presumably frozen diamictite down to almost two meters' depth. Similar late Precambrian sand wedges have been reported from Australia, Scotland, and Spitzbergen, while modern ones are well known not only on earth but also on Mars. The myriad periglacial phenomena in Mauritania are classic and indisputable exemplars of glacial geology. They show that once local glaciers had finished feeding icebergs to the sea, North Africa experienced a spell of cold and arid climate that lasted hundreds, or even thousands, of years, before rising sea levels eventually drowned the tundra landscape, preserving it faithfully beneath a sedimentary drape.

Despite such apparently incontrovertible evidence, criticism of the glacial hypothesis did not stop in 1976 but was repeated for each and every occurrence of glacial diamictite of similar age across the world. Geology, being a historical science, has come in for a fair amount of criticism over the years, unfairly I think, for its emphasis, not on fundamental laws, but on reconstructing past events that cannot be witnessed and so are untestable in the strictest sense. I consider this unfair because geology, like all historically based study, applies a forensic eye, which allows us to frame testable hypotheses that after further observation and experiment can be refuted or refined. In my experience, getting geologists to agree on a precise interpretation for any given outcrop is extremely challenging and requires an almost fanatical devotion to making the case as watertight as possible, often by taking thousands of measurements, mapping outcrop patterns in tremendous detail, or repeating innumerable geochemical or geophysical tests. Geologists, like lawyers, build up their case on circumstantial evidence. There is never a confession and seldom a smoking gun, but juries of scientists the world over ponder the evidence using the balance of probability. A good example of this comes from another striated pavement of similar age, this time from Norway.

The very first report of a glacially striated pavement of late Precambrian age was made by Hans Henrik Reusch, who was the director of the Norwegian Geological Survey from 1888 until 1921. During that tenure, in 1891, he drew attention to glacial striations (not uncommon in Norway) that seemed to derive "from a period much older than the ice age." Just as in Mauritania, the polished, scratched, and grooved pavement at Varangerfjørd was overlain by a diamictite, and according to some geologists represented an ancient surface of erosion (called an unconformity) across a large part of Norway. During the intervening years, more than forty scientific papers have been published on this surface alone with diverse interpretations. Some authors accept the glacial evidence but consider it to relate to more recent ice ages, while others prefer an entirely non-glacial origin. This is not a case of interpretations approaching a consensus once more evidence can be brought to bear. In fact, the papers from the non-glacial school have come to challenge, if not dominate, the field. Most troublingly, even Brian Harland, that most indomitable proponent of the Great Infra-Cambrian Glaciation, was a skeptic in this particular instance. It is laughably easy to cherry-pick such disagreement in support of an alternative to all Precambrian glaciation, but that would be wrong. The glacial hypothesis, and indeed any geological hypothesis, does not succeed or fail due to the interpretation of one particular outcrop. It is the combined weight of contextual evidence that lets it sink or swim.[6]

Four years after first introducing me to Mauritania, Max Deynoux was part of a nationwide French research project to study the late Precambrian glaciation. This was one of many such projects around the world spawned by the excitement generated by the resurgent Snowball Earth hypothesis, about which we will hear more later. The project was led by Anne Nédélec, a geophysicist from Toulouse who wanted to find out the latitudes to which ice sheets had once extended. And so, in 2000, I was delighted to receive another invitation to join Max on an adventure through northwest Africa, this time crossing borders from Senegal into Mali. It was to be a memorable trip, although it began inauspiciously in a seedy hotel in Dakar on the ocean shore. My jet lag caused me to sleep soundly through the night, only vaguely aware of the crashing of the waves through my window, stupidly left ajar to feel the cool sea breeze. I awoke only to find both my eyes swollen shut by mosquito bites.

Throughout the trip, we were plagued by insects. Mosquitoes seemed to know as if by magic which part of my body was peeking out of the sleeping bag, and tsetse flies proved equally persistent during the heat of the day. We were in Senegal to look for a certain valley near Kédougou, a city close to the country's southeastern border. Max had been there just once before, and that had been a long time ago, only shortly after Senegal's independence from France. Surprisingly, it was a British group that had brought the area to the attention of the scientific community. In geology, easily recognizable packages of rock that can be represented on a map are called formations, which are generally named after the place where they were first defined. The better part of our first day was spent looking for the village of Hassanah Diallo, the name given to this particular formation by the pioneering British team. After being pointed from village to village, we were finally introduced to a gentleman who remembered the geologists in question and could show us where they had worked. It turned out that Hassanah Diallo was not in fact the name of the village but the name of its chief, and both he and his village tended to move around. Born in neighboring Tanaga, he was now the chief of the sprawling village of Pelel Kindessa. Fortunately for us, Mr. Diallo had not moved very far, and was all smiles when we eventually tracked him down.

The reason we had come to Walidiala Valley was to address full-on the concerns of skeptics like Schermerhorn (whose full chosen name was Lodewijk Jacob Gerard Carel Eduard, quite a mouthful, though he usually shortened it to LJG). LJG's chief objection was that it was hard to distinguish marine tillites from debris flow deposits formed by submarine landslips,

and that today they are often one and the same as the passage
of ice caps over continental margins can lead to slope instability, resulting in hugely thick successions of mixed-up mud and
rock. Max's painstaking measurements in the same region, the
Taoudeni Basin, but much farther to the north in Mauritania
and Algeria, had shown that boulders had been transported
from north to south, but he had mostly focused on terrestrial
glacial deposits like till or moraine. However, ice sheets flow
downhill, presumably toward the sea, and so we presumed that
somewhere to the south of the basin we ought to find evidence
that glaciers had once met the sea. In those few days, I learned
a lot about the various kinds of sedimentary conglomerate one
might find at the ocean's margins, and just why it is so hard to
tell glaciogenic from non-glaciogenic diamictites. In the Walidiala Valley, many of the outcrops we looked at were of diamictite, but none showed any definitive sign of glacial origins,
with one exception.

"Dropstone" is a term you need to be very wary of. Sedimentologists learn to hate such words, because although they
are undoubted interpretations, they are often used interchange-
ably with their baggage-free and purely descriptive alternatives.
Therefore, dropstone is often synonymized with lonestone, and
diamictite confused with tillite, and so on. I could go on but
will desist, as this is a topic that can become really tedious, so
let's get straight to the point. At any one outcrop, it is often
nigh impossible to determine whether a lone pebble in a fine-
grained mudrock is truly a stone that plunged vertically into
the sediment—that is, a dropstone, and as such a plausible can-
didate for ice-rafted debris. This ambiguity is unsurprising, as
large boulders can be transported on the tops of mud flows,
while later compaction, during the transformation from mud
to rock, will lend any outsized clasts the superficial appearance
of having dropped into a muddy sediment. In such cases, the

Figure 3. Ice-rafted dropstone. Max Deynoux measuring a potential
glacial dropstone (Tanague, Walidiala Valley, Senegal). True
dropstones pierce underlying layers and disrupt neighboring layers.
By contrast, overlying layers merely drape over the outsized stone.
(Photo by author, 2002)

highly compressible mud layers will wrap around incompress-
ible clasts to end up looking for all the world like a dropstone.
To an expert eye, however, some lonestones can be confidently
interpreted as dropstones by the fact that they pierce and deform the existing laminae of the mud layers before compaction,
while being draped by later layers. It was just such a dropstone
horizon that we found at Walidiala and at precisely the same
stratigraphic level as the Mauritanian till (Figure 3). We had
found the seafloor that neighbored the Taoudeni ice shelf 635
million years ago.

     As Harland maintained, although diamictites are ubiqui-
tous in rock successions of late Precambrian age, dropstones
are often the only unambiguous evidence for glaciation in those
diamictites. Some dropstones are faceted and striated by ice

action, but this is rare because many boulders locked in icebergs were originally transported as glacial outwash, and so have been rounded by rivers fed by melting ice caps and glaciers. More commonly, glacial dropstones cluster in groups, a sure sign that they were deposited still wedged inside a block of ice that subsequently melted on the seafloor. And occasionally, dropstones are themselves composed of unlithified moraine (till pellets) that could never have survived the rough-and-tumble of a submarine landslip. Geologists gather together many such suggestive clues, however small and apparently insignificant, to build up a four-dimensional picture (3D + time) of the earth's past environment and, in this case, to trace the waxing and waning of ice sheets.

Each day we climbed the steep hill, walking through time from the oldest glacially affected rocks at the base of the valley to the youngest at the top that showed no signs of the action of ice. This marine equivalent of the Mauritanian tillites was much thicker, as might be expected, and comprised multiple debris flows all scouring into each other and only occasionally separated by an undisturbed layer with dropstones. It was here that, from time to time, the aforementioned baboons would race across the valley in a remarkably successful attempt to intimidate me by throwing rocks on my head from the cliffs above. On one day, we walked all the way to the top, where we discovered that the Fulani (or Pel) people who inhabited the high, flat plateaus there were wholly different from the darker Wolof people of the valleys. The Fulani are the world's largest nomadic, pastoral group and stretch across Africa from west to east driving cattle, goats, and sheep, and selling their milk products. It was a weird feeling appearing unannounced and seemingly out of nowhere to walk conspicuously through the middle of their village. It was even stranger to discover much later that—in walking into their village—we had crossed from Senegal into

Guinea and back again. The nomadic existence of Saharan Africans shows no respect for national borders. Or perhaps it would be more accurate to say that the colonial borders of Saharan Africa show no respect for the nomadic lifestyles of the people who live there.

From Senegal we wanted to cross the border into Mali. Max knew that there would likely be no official at the Kéniéba border crossing, which was just a ford, only negotiable in the dry season, so we located a policeman who was having breakfast in Kédougou and persuaded him to come along with us to stamp our passports. Pascal Affaton, a seasoned geologist originally from Benin, amused himself by terrifying the poor man, who was unused to sleeping wild, with tales of lions, snakes, and crocodiles when we made camp that night. We set up a proper camp each evening before the cooking began and each morning took down the pole tent and folded the tables and chairs. It was all surprisingly civilized. When we finally reached the border, what struck me most were the small children, toddlers mostly and young infants, with bellies so distended that their belly buttons stuck out like tubes. The Ghanaians call this "the sickness the baby gets when the new baby comes," or Kwashiorkor, and it is apparently caused by the all-too-common protein deficiency that children get once they have been weaned. Such malnutrition does not lead to emaciation because the young infants may still consume sufficient calories, but the lack of protein causes swelling and water retention. The Kéniéba area has been a gold mining area for hundreds of years, but few families find their fortunes there. We stopped on occasion to watch the older boys climb down as far as thirty meters inside tiny rectangular shafts, while their parents and siblings panned through the sediment they brought up. Once panned, the gold flakes are purified by mixing with toxic mercury, which is then burned off. Artisanal mining throughout this region leads to chronic mercury poisoning on top of the more general ills of child labor. It was a sad sight to witness, which, judging from recent reports, has yet to change for the better.

Our journey through western Mali to the city of Kayes and beyond could not have been more different. It was not long before we were camping alongside one of the most impressive waterfalls in the world, the Gouina Falls, sometimes called the Niagara of Mali. Despite its beauty, we did not need to share the sight with anyone, apart from the occasional rally driver disturbing the serenity by roaring back after finishing the Paris-to-Dakar rally. On the way to the falls, Max occasionally pointed out aspects of the geology. We had passed dense, blood-red rocks that his student Jean-Noel Proust had discovered were deposited on the seafloor, either adjacent to the great ice caps or perhaps even beneath floating ice. Although most commonly associated with much older times in earth history, the brief reappearance of iron-rich deposits, called banded iron formations, during the early stages of the late Precambrian glaciation was to play a key role in the Snowball Earth debate to come. However, we had not come to see the iron formations or even the abundant diamictites. We were in Mali to look at how the ice age ended.

After we had raced for hours along an almost perfectly flat desert floor, passing baobab trees and occasional huts, but not much else, we came upon a series of inselbergs, those isolated mountains in the desert that appear to have resisted the incessant bite of time and the elements. The rocky islands near Kayes are made up of reddish sandstone, once sand, piled high in vast migrating dunes that were then planed by erosion before being overlain by another set of megadunes that migrated in a slightly different, but still southward direction. The planar surface separating these two dune sets is called a supersurface and represents the beginning of a new cycle of erosion and deposition. Max invited us to climb the inselberg where Jean-Noel and he had discovered the same polygonal cracks as in Mauritania, more than a thousand kilometers to the south, underlying each supersurface. As the late Precambrian glaciation waned, much of northwest Africa had been planed into an almost perfectly horizontal, barren desert of windblown sand and permafrost over an area of more than a million square kilometers.[7]

The presence of ice greatly modifies patterns of atmospheric circulation, so it comes as no surprise that aeolian deposits were a feature of the late Precambrian periglacial landscape, just as they are today. Winds near glacial margins are often strong and persistent because of the cold air that sinks off ice caps, redistributing glacial flour as loess and sand grains into immense dune fields. A good example of a periglacial sand dunescape today is the Sand Hills region of Nebraska, which covers an area of 60,000 square kilometers. It was a large, active dune only 10,000 years ago, at the end of the last ice age, but is today largely stabilized by grass.

The visit to Mali completed a picture for me of glaciation in northwest Africa, painstakingly built up over decades of research by teams of experts. I had seen how the glaciers had scoured the uplifted continent, transporting debris from the highlands of the north to the southern plains, across which vast dunes migrated, driven by the prevailing wind. I had seen how some of those ice sheets had ploughed the continental shelves, dredging them like gigantic diggers, causing submarine landslides. And I had seen how floating icebergs had occasionally dropped their cargoes into this mixed-up substrate. In total, I spent only forty days in Africa, during three separate trips to four different countries. That may sound like a lot, but I am acutely aware that I was extremely fortunate to be shown at breakneck speed some of Africa's geological gems. I am awestruck by how much human endeavor has led to even this much knowledge about the late Precambrian glaciation in just one region of our planet. It cannot be measured in hours, or even days, of work. If anything, one might have to measure it in lifetimes or careers. Even Max's work that I would happily label pioneering had been preceded by others who themselves had spent many years producing the first topographic maps, aerial photographs, and geological surveys that would form the basis of all future work. The accumulation of knowledge through meticulous observation is an important aspect of science, so rarely feted, but without which no real progress can be made.

Despite such strides, many questions still needed answering. How long did this ice age last? How extensive was it? Could this in any way be described as Harland's "Great Infra-Cambrian Glaciation," or was it as Schermerhorn believed, merely restricted to mountainous regions and high latitudes? To answer such questions, scientists needed to find out the age of these deposits, their precise locations so long ago, and how they fit in with the other, frequently less convincing reports of glaciation from all over the world. All this was uncertain until quite recently, but there were already some clues.

For years, it had been known that one thin bed of rock could be found almost everywhere throughout northwest Africa. This rock layer overlies the striated pavements and diamictites of Mauritania. It overlies the debris flows and dropstone-laden mudrocks of Senegal. And it overlies the windswept tundra landscapes of Mali. Such marker beds are crucial to geologists, but none is as widespread as this one. Indeed, this rock may well be the only known example of a globally deposited sediment in all earth history, and one that, once persuaded to give up its secrets, would hold the key to the truly exceptional nature of the late Precambrian glaciation. But we are getting ahead of ourselves. Let me first introduce you to another sunburned but eerily glaciated continent, Australia.

3

Meltwater Plume

About 635 million years ago, at the end of the Cryogenian ice age, a layer of carbonate rock formed on top of glacial landscapes worldwide in the only global depositional event in the history of the planet. There are many indications that this "cap dolostone" was deposited during deglaciation. Forensic examination of these unique rocks suggests that abrupt melting sealed the cold, dense glacial oceans beneath a thick surface meltwater plume for up to a hundred thousand years.

Several years after my first desert encounter with Precambrian glaciation, I was again racing over scorched terrain in a weather-beaten Toyota. In the intervening years I had belatedly taken my driving test, so now it was me behind the wheel far from anything remotely like a road. As one member of a project of the UNESCO International Geoscience Programme, or IGCP, it was my task to guide a group of fellow rock enthusiasts through Australia's red center. I had promised them that we would first stop in South Australia's Flinders Ranges to catch a glimpse of what is, for geologists, one of the most famous rock outcrops in the world. Camping each night in the outback, largely barren except for rare oases of squawking cockatoos, our friendly group shared songs by the light of the dying embers and hearty breakfasts at the crack of dawn. We were geologists from China, Korea, the United States, Canada, France, and Australia, brought together by our shared fascination with the earth's deep past.

The cream-colored rock layers at Enorama Creek may look unprepossessing, if not downright drab, but they are unquestionably important. In 2003, the International Commission on Stratigraphy, a formal body made up of world experts of this time interval, voted for this "Nuccaleena Formation" to mark the official start of the final chapter of Precambrian time. Since that momentous occasion, rocks above the base of the Nuccaleena Formation have been assigned to the Ediacaran Period, while rocks below now belong to the Cryogenian Period, so named because of its freezing cold climate. This crucial juncture in earth history is marked by a shiny brass disc, which shows the precise level of the GSSP, or "Global Boundary Stratotype Section and Point." GSSPs are decided upon after many years and even decades of study in a process that can sometimes be marred by petty wrangling, fueled more by national pride than scientific evidence. Geologists from all countries use such "golden spikes" to subdivide time and correlate strata, or rock layers, around the world. At the Enorama Creek GSSP locality, the golden spike was hammered in to recall the catastrophically abrupt conclusion to the great late Precambrian ice age (Figure 4). It also marks the beginning of a new biosphere, one in which animals would increasingly come to dominate.[1]

The so-called cap dolostones of Australia are well known marker beds that drape, or "cap," glacial diamictites. Being so

Figure 4. Basal Ediacaran golden spike. The author at
the internationally agreed-upon starting point of the Ediacaran
Period (the *golden spike*), Enorama Creek, Flinders Ranges, South
Australia. The buff-colored "cap dolostone" beyond the information
panel marks the formal base, which directly overlies glacially
influenced, red diamictite of the Cryogenian Period.
(Photo by Markus Sachse, 2005)

ubiquitous, they have been used for many years to correlate
sedimentary layering across a vast area, stretching almost 4,000
kilometers from Tasmania in the south to the Kimberley Ranges
in the north. Visually, they are almost entirely unremarkable,
being generally less than ten meters thick, and composed of
dolomite, which is a common marine carbonate mineral, comprising calcium and magnesium cations in roughly equal measure. There are some odd features that we will take a look at in
due course, but I think it needs to be emphasized that this is a
rock that would be very easy to overlook were it not for its in-

credibly vast, even global, extent. Sediment types vary greatly from place to place in the world's oceans today, reflecting the diverse environments in which they were deposited, so how could a marine sedimentary environment remain constant throughout an entire region, let alone the whole world? This riddle remained unsolved for many years.

It was the Australian Geological Survey that first recognized the importance of cap dolostones for regional correlation, following in the footsteps of Douglas Mawson. Shortly before our trip set out, an elderly gentleman by the name of Bob Dalgarno asked through a mutual acquaintance whether he could tag along, and I passed on my acquiescence without giving it much thought. To my amazement and not a little embarrassment I soon discovered that it was Bob who had led the first detailed geological surveys of the Flinders Ranges. The moment it really struck home was when he showed me an old black-and-white photo of Enorama Creek that he had taken as a young field geologist. Not much had changed since, not even the eucalyptus tree, which had grown interminably slowly in the intervening forty-five years. With characteristic charm and humility, Bob explained to the group that he had always guessed at a wider significance for the strangely ubiquitous cap dolostone, to which he gave the name Nuccaleena Formation, but it was not his job to speculate. No sooner had he finished his work there than he was moved on to somewhere else to make the next map.

Bob is another one of those unsung heroes who have contributed careers of time and energy to the accumulation of geological knowledge, and without whom our grand hypotheses would have no substance. Similar surveys were carried out in northwest Africa around the same time, and with the same conclusions. The French geological teams of the 1950s and 1960s came up with the concept of the "triad," referring to how a thin dolostone layer could always be found sandwiched between glacially influenced rocks below and fine-grained mudstones above. They were able to trace the triad not only across the Taoudeni Basin in present-day Mauritania, Algeria, Mali, and Senegal, but also into the neighboring Volta Basin of Benin, Togo, Burkina Faso, and Niger. The cap dolostone, and indeed the triad, has now been correlated throughout the world, everywhere draping glacially influenced landscapes and seafloor of Cryogenian age. Despite their unspectacular appearance, the basal Ediacaran cap dolostone is the only rock of global extent known to exist, and that fact alone makes it truly exceptional.

So, what exactly is a cap dolostone? In short, it is a layered sedimentary rock, composed of a mineral called dolomite, with the approximate chemical formula $CaMg(CO_3)_2$. Cap dolostones average less than twenty meters in thickness and sit with knife-sharp contact above a diverse range of glacially influenced rocks all over the world. Cap dolostones have been reported from twenty different continental fragments, called cratons, across the globe. Indeed, wherever rock successions of this age have been investigated they have never been wholly absent. The underlying diamictites can represent both terrestrial and marine environments: from submerged tundra landscapes like those in Mauritania to the sorts of debris-laden seafloor we can find in Senegal. Although reports have been made of the occasional ice-rafted dropstone within their lowermost layers, cap dolostones and overlying strata show no sign of local glacial influence. The widespread distribution of cap dolostones identifies them as what geologists call *transgressive* deposits, reflecting a significant rise in sea level that must have submerged vast areas of what—during glacial times—had been dry land.

Most of us are used to the notion that sea level can undergo a global, or *eustatic,* sea level rise as a result of deglaciation. At present, you might be concerned that the melting of ice sheets and glaciers could raise sea level by a few tens of centimeters during your lifetime, perhaps enough to drown some low-lying island atolls in the Pacific Ocean. In comparison, if both of the world's major ice caps in Greenland and Antarctica were to melt away, something that might take a thousand years or more, it is estimated that sea levels would rise by a whopping seventy meters, drowning coastal cities across the world. During the last glacial maximum, only 20,000 years ago, sea levels were lower by as much as 125 meters, and even as recently as 11,000 years ago, beaches were 110 meters lower than today. It must have been truly shocking to witness the inexorable submergence of coastal landscapes around the world as many of our recent ancestors must have done. Some people even attribute the surprisingly large number of flood myths around the world, including the most famous example of Noah's flood, to the relentless rise of the sea by as much as six centimeters per year between 19,000 and 8,000 years ago. This may sound like a big deal, and it is, but at the end of the Cryogenian ice age, sea levels are thought to have risen over an order of magnitude higher and faster.

Despite recent deglaciation, we still live in glacial times, and icebergs still roam the oceans, dropping stones onto muddy seafloors around the world. Indeed, the planet has experienced the waxing and waning of ice caps for millions of years, with the result that sea levels have fallen and risen every 100,000 years or so. Such cyclical behavior is typical of the earth's climate history, whether ice was present or not, and this is due to the various wobbles and eccentricities in our planet's orbit around the sun. However, the lack of significant ice-rafted debris in cap dolostones at any latitude reflects how the melting at the end of the Cryogenian ice age was more abrupt, more profound, and more final than anything experienced before or since. Once melting began, it seems to have been a runaway, unstoppable process, and the deposition of marine strata on top of extensive terrestrial landscapes illustrates this in a qualitative way. Although I prefer to quantify change, this is understandably challenging when dealing with an event hundreds of millions of years in the past. Thankfully, clues can still be found if we know where to look.

Between my African trips with Max Deynoux, I returned to Mauritania with a different IGCP group. Our purpose was to study the older rocks that made up the glacially scoured surface beneath the Cryogenian diamictites. On occasions, I was able to separate off from the rest of the group to take another look at the cap dolostones that had so puzzled me on my previous trip. Since that time, the Sahara has become a dangerous place, where it is easy to be taken hostage or disappear without a trace. But back then, I felt very differently. I remember well how the local women and children invited me with knowing glances to stroke a baby hedgehog. Obligingly, I did so, and its teeth latched onto my finger so hard that no amount of coaxing could remove it. Clearly, that had been the plan all along, and great merriment ensued all around. It was a fun trip, and pleasantly easygoing. This time I flew to Atar directly from Marseilles, landing right on top of the very rocks we were there to study, and with immediate views of the line of cap dolostones in the Falaise d'Atar. My taxi experience of just a few years earlier now seemed a world away.

At Jbéliat, the cap dolostone Max had shown me was only a few meters thick, but with such disrupted and contorted layering that it was a challenge even to measure its true thickness. Thin layers of dolostone were cemented together by dolomite crystals, which lined "sheet" cracks between the bedding that in places had buckled the layers into pointed tepees, a sure sign that cementation, the process that turns soft sediment into hard rock, had taken place during or only shortly after deposition. Balls of barite, which is barium sulfate in mineral form, were strewn over the cap dolostone with apparent abandon, and even Max could make neither head nor tail of why they were there. Overlying the entire thin package, a bizarre layer of dark purple limestone (calcium carbonate) just thirty centimeters thick only exacerbated the enigma. This time around, I could look at the outcrops with more experienced eyes and began to wonder whether the cap dolostone had really been deposited during a simple one-step sea level rise.

On my return, I wanted to pay more attention to the rare, thicker cap deposits that seemed to fill troughs, possibly scoured into the rocky basement by rivers of meltwater. At Guelb Nouatil there are, unusually, two levels of thin dolostone rather than just the one, and these were separated by as much as thirty meters of sandstone, "cross-bedded" like Saharan dunes, with only the upper level buckled by cementation. Intriguingly, the pointed tops of the tepees had been leveled flat by erosion, while the same thin, purple limestone lay directly on top of the erosional surface. A close inspection of the upper dolostone shows it to be riddled with small caves, where the carbonate rock had apparently been dissolved away by rainwater. These karst caves must have formed early on because their interiors were coated with barite, with any remaining gaps filled in by the enigmatic purple limestone. Occasionally, if you are lucky, you can find one outcrop that can bring all hitherto puzzling aspects of the geology into sharper focus; Guelb Nouatil is one such place. It seemed now plausible that the deglacial sea level rise had proceeded in two stages, with the erosional surface marking a period when the dolostone had briefly emerged from beneath the waves to be ravaged by the elements.[2]

This tale of two transgressions, a technical term for when sedimentary environments migrate inland due to rising sea level, can be seen the world over. A couple of years later I was intrigued to read of a similar account from the Mackenzie Mountains, high up in the Canadian Rockies. The three authors, Noel James, Guy Narbonne, and Kurt Kyser, all seasoned Precambrian experts from Kingston University in Ontario, described similarly truncated tepees. They even went so far as to predict the existence of karstic caves that, if present, would, they speculated, prove their hypothesis of a two-stage transgression. In Canada, however, they were unable to find any, as the caves unfortunately would have been buried far beneath the mountain scree. Other cap dolostones from around the world show the same signs of emergence, but what can this mean? Sea level must have risen twice, yes, presumably due to melting ice caps, but also must have fallen again between those two stages. Yet there is no evidence for any return to glaciation that might have caused a drop in eustatic sea level. If we can rule out global sea level fall, then an intriguing, but almost incredible, possibility presents itself, which only a closer look at the rocks can tease out.[3]

Before we get to that, we need to take a slight diversion. You may recall that the interior of our planet resembles an onion. If you were to cut it in two with a sharp knife, its layering would be immediately obvious, showing a metallic core at the center, overlain by layers of silicate rock that decrease in density toward its edge, where finally the lightest layer, the crust, forms the scum at the surface. The crust has especial importance for us because we live on top of it. We humans have uncovered our planet's layering chiefly by measuring the time it takes for seismic waves after earthquakes or explosions to refract through or rebound from those layers. Through much hard work and detailed study, the earth's innermost secrets have now been laid bare, until their gross detail can no longer be disputed. We now know that many of the earth's surface phenomena, from continental drift to mountain building, can be explained by the fact that although the earth's outermost layers, which we call the lithosphere, are rigid and brittle, deeper parts of the mantle act more like a slow-moving liquid due to the high temperatures and pressures in the earth's interior.

Liquids don't break under pressure. Instead, material moves down and sideways to compensate, and this is just what happens when ice caps thicken and thin as they have done many times over the past few million years. Because ice has less than a third of the density of mantle rock, roughly speaking, it takes just over three kilometers of packed ice, about the same as currently exists in the ice caps of Greenland and Antarctica, to cause the earth's crust to flex downward by a whopping one kilometer into the underlying mantle. For this reason, much of Greenland's ice-covered land has sunk beneath sea level, and if all the world's ice caps melted away over the next century, more than half of Greenland would become ocean. It would cease to look like the Greenland familiar to us from maps of the world, but for a time would become a chain of islands and archipelagoes, much like the modern Philippines. But only for a time. Just as extra weight will force the crust to sink into the mantle, lifting off that weight will cause the land surface to bounce back, more or less to where it was before. This phenomenon is well understood and is called isostatic rebound. Sea level changes caused by isostasy vary from region to region, distinguishing them from the eustatic rises and falls caused directly by deglaciation and glaciation, respectively.

No sooner had the glaciers left northern Europe and America than the land began to relax, bouncing back to where it had been more than a hundred thousand years earlier. However, isostatic rebound is a slow process. Although all eustatic sea level rise ended more than five thousand years ago, isostatic sea level change lags far behind and is still occurring today.

Indeed, parts of the northern hemisphere, such as the Gulf of Bothnia in Scandinavia, are still rising in places by more than a centimeter every year. Raised beaches tell us that many of these areas have risen by more than 300 meters in total, which suggests that the ice sheets there reached at least a kilometer in thickness. Isostatic rebound begins quickly, then decreases exponentially until equilibrium is reestablished after many thousands of years. Just as some higher latitude regions, like Scotland, are gaining in stature, adjacent regions, like England, are sinking, engaged in a game of glacial seesaw around a central pivot. If you have ever wondered why the beaches of the Great Lakes are more splendid on the Canadian side, this is because the beaches on the American side are slowly sinking, while the Canadian ones ride high.

Once the burden of ice has been removed from an area, much of the rebound takes place as an elastic response to pressure release. However, rebound continues over a longer interval as mantle material flows slowly from the ocean basins into the areas beneath once glaciated continents. This "isostatic adjustment," together with the added burden of more water, causes ocean basins to sink toward the gravitational center of the earth and away from the continents, now made less massive by the melting of their burden of ice. When averaged over the entire ocean, the effects of the last deglaciation can still be detected in the almost imperceptibly slow sinking of the seafloor by about three centimeters per century.

You may be forgiven for wondering what all this has to do with the Cryogenian ice age. To answer that, let's return to our intriguing cap dolostones. As we saw earlier, glacial landscapes worldwide show clear signs of having been submerged by the rising meltwaters. In many places, overlying cap dolostones were then somehow exposed to the elements before being submerged again, this time for good. This can most easily be explained as a result of the complex interplay between the eustatic rise of the ocean caused by deglaciation and the isostatic rise of the land caused by the local removal of thick ice sheets. After the glaciers retreated, both the land and the sea rose. Initially, the sea rose more quickly than the land, but as in the story of the tortoise and the hare, the land eventually caught up, stranding the once submarine cap dolostone over vast expanses of low-lying continent. We will deal with how sea levels continued to rise in a moment, but it is worth reflecting on what this timing means for the ubiquitous cap dolostones. Evidence for exposure is generally found at or close to the tops of these enigmatic rocks, which vary in thickness, but average about eighteen meters overall. It seems that most ice caps had melted before dolostone deposition, while isostatic adjustment, although slow, would largely have been completed within 10,000 years of deglaciation. A quick back-of-the-envelope calculation gives us a rough idea that cap dolostones formed at rates of about one millimeter to ten centimeters per year, as a bare minimum.

Coincidentally, these rates are also typical of the slowest- and fastest-growing corals, respectively, which are also made of carbonate. But that is where the comparison breaks down. The fastest-growing corals live in very shallow, well-lit, and warm coastal areas, where seawater is highly oversaturated with respect to calcium carbonate; all carbonate minerals become less soluble at higher temperatures. However, cap dolostones covered far greater areas of the world's continental shelves and formed in deep shelf as well as shallow marine environments. They are densely layered deposits, not at all analogous to the fast-growing but spindly, antler-like arms of modern hard (scleractinian) corals. The rapidity and vast extent of this carbonate precipitation event dispute a well-regarded hypothesis that cap dolostones represent drowned environments that suddenly find themselves far out to sea when sea levels rise. Such "high-stand deposits" often contain a fair amount of precipitated carbonate, or cemented "hardground," because the river deltas that would normally supply detritus have been drowned. However, due to the lack of any other sedimentary material, such condensed deposits form incredibly slowly, gaining mere centimeters in millions of years.

The global triad of diamictite-dolostone-mudstone suggests that sea levels continued to rise even once isostatic adjustment had fully ended. If correct, then the base of this second transgression could represent just such a condensed, high-stand, hardground deposit, and it was with this in mind that I took a longer look at the purple limestone that filled the cracks, crevices, and tops of the "drowned" cap dolostone at Guelb Nouatil. To the naked eye, the limestone seems to consist of dark flakes, nestled against each other in what sedimentologists would call an edgewise breccia, cemented together by calcium carbonate, but there are two slightly surprising aspects to this deposit. First, the rock contains exquisite, elongated crystal blades of barite ($BaSO_4$). You may remember that the rocks of neighboring Jbéliat also contained barite, which is known to be a feature of the tops of cap dolostones in other places, too, such as Australia, Canada, Norway, and China. This is a key observation that we will return to again below. The second odd feature is that the flakes have a distinctly volcanic look about them. Their mineral composition and textures show that they were once incandescent blobs of melted rock that were then quenched to glassy shards, presumably after being thrown out of a volcano. Volcanic eruptions occurred at this time right across northwestern Africa as well as Brazil, its then nearest neighbor, although it is not immediately clear why. After much head-scratching, volcanism at the start of the Ediacaran Period is now widely attributed to an abrupt release of overburden pressure following the rapid melting of thick ice sheets.

Unlike the cap dolostones, the overlying limestone unit does not contain much in the way of terrestrial detritus, which strongly supports the idea that the transgressive cap dolostones were in turn drowned by continued sea level rise, seemingly well after ice sheets had retreated inland of the coastal regions. It has been suggested that some of this second sea level rise could be due to the melting of inland ice caps and glaciated mountain regions elsewhere in the world. Continuing deglaciation, together with smaller effects caused by the thermal expansion of liquid water, is needed to explain the magnitude of this second relative sea level rise, which suggests an extraordinary possibility. Not only were cap dolostones formed during isostatic adjustment, a period of less than 10,000 years, but during deglaciation itself, so their deposition could have been even more rapid than we estimated above. What evidence is there for such rapid accumulation?

Although I wrote above that cap dolostones are unremarkable, several studies have highlighted unusual features in addition to the sheet-crack cements and tepee-like structures that I mentioned already. Chief among them are stromatolites. Stromatolites are domes of wrinkly laminations formed by layer upon layer of microbial carpet, with photosynthetic cyanobacteria forming the top layer, where they can best make use of the sun's rays. They can still be found today off the west coast of Australia, but the largest and best-known forms are all Precambrian in age. Although there were no corals and therefore no coral reefs at those times, stromatolite reefs were common. Our international group was in Mauritania at the invitation of the stromatolite expert Janine Sarfati, who had spent her career studying the immense underwater forests of narrow pyramid-shaped reefs called Conophyton. The airport in Atar is built upon one such reef, and I recall groups of children sitting on the airport runway eagerly selling fist-sized blocks of stromatolite as novelty paper weights. Despite the absence of fish, diving the pre-Cryogenian tropical reefs with their branching Conophyton trunks and boughs would still have been an amazing experience, with colorful seaweeds and blue-green algae (cyanobacteria) along with a plethora of weird and wonderful single-celled protists, many living symbiotically or parasitically—depending on your point of view—off the energy provided by their photosynthetic neighbors.

Most stromatolites, by their nature, form in shallow waters. Too deep, and the sun's rays would have no effect. The cap dolostone stromatolites are unique in their appearance, forming almost impossibly thin towers of laminated dolomite, supported by early dolomite cement in between. It seems that they had to race to keep up with the rising sea levels, losing the battle in deeper, more distal realms, where there is only reworked dolomite, but just about managing to keep up where conditions were perfect, as in upper-slope areas where cap dolostones can attain more than one hundred meters' thickness, made up almost entirely of so-called "tubestone" reefs. Slowly but surely, we are building up a picture of deglaciation unlike any the world has ever seen. As sea levels rose, dolomite began to be deposited farther and farther inland, but all within the several-thousand-year time frame permissible from isostatic adjustment. A huge volume of meltwater is implied, vastly greater than released since the last glacial maximum, and with inevitable consequences that we are now beginning to recognize. Some further clues can be found in the mineralogy.[4]

Barite is a curious mineral. One of the densest minerals known, it is also one of the most insoluble, so insoluble in fact that its cation barium (Ba^{2+}) can scarcely coexist in solution with its companion anion sulfate (SO_4^{2-}) in aqueous environments anywhere on earth. To form at all, in any sizable amount, fluids that contain barium (but no sulfate) need to meet fluids that contain sulfate (but no barium). When they meet, barite precipitation will be the inevitable and immediate result. Seawater contains lots of sulfate, as it is the most abundant ion in the ocean after sodium and chloride, the main constituents of table salt. Because sulfate is so abundant in today's oceans, barite generally forms only within the pore waters of an already deposited sediment, where the actions of sulfate-reducing bacteria have removed sulfate entirely. In these dark, organically laden sediments, barium ions can build up at depth, allowing a barite mineral front to form just at the boundary between methane-producing bacteria below and sulfate-reducing bacteria above. The barite front continually redissolves and reprecipitates, moving upward to keep pace with sedimentation. This is the usual case, but cap dolostone barite does not resemble this scenario at all.

Barite domes and cements in Mauritania formed rapidly in karstic caves that were drowned by rising seas only shortly after the caves had formed. At other localities, like in the Mackenzie Mountains of Canada, barite formed directly on the seafloor together with aragonite, a type of calcium carbonate mineral, in arrays of crystal fans rooted to the tops of cap dolostones. In Canada, barite fans are associated with iron-stained dolostone for hundreds of kilometers, and the same association can be found in far-off Mongolia. This is a genuine puzzle, for how can barite grow directly in seawater if barium and sulfate ions can barely coexist? The presence of barite across the world can only be explained by the mixing of two very distinct waters, one rich in sulfate and the other devoid of sulfate, and most likely deficient in oxygen, too. Like the cap dolostones and the tubestone stromatolites, this barite precipitation event seems to have been a unique occurrence in earth history that must relate to the extraordinary conditions at the end of Cryogenian glaciation. But how?

In 2004, a new Hollywood film, *The Day After Tomorrow*, made its way through cinemas, becoming the sixth highest grossing picture of that year. It portrayed a natural catastrophe that would threaten civilization itself. Having now finally seen the film myself, I am not sure whether the greater disaster was the contrived plot or the supposed event that set the story in motion. However, leaving the film's quality aside, as with many such disaster movies there is an inkling of truth, however small. In the film, Jack the paleoclimatologist does the movie scientist's job of warning everyone about the impending doom soon to be wreaked upon the world by melting ice. According to Jack's theory, which predictably falls on deaf ears, polar meltwaters are spreading around the globe, forming a freshwater lens that will disrupt ocean circulation patterns and global climate. He predicts, and is proven right (in the film), that the vertical movement of water bodies would totally cease in the Atlantic Ocean, causing a return to glaciation almost overnight. The theory stands up in the sense that freshwater is undoubtedly less dense than salty seawater and so might be expected to "float" upon it for a time, at least until the wind and waves mix it all up again. Moreover, as the surface layer gradually warmed up from the heat of the sun, it would become even less dense through thermal expansion, thus exacerbating the buoyancy contrast. So far so good. However, the scientific veracity of the plot soon goes belly up.

A recent ocean modeling exercise, reportedly inspired by said film, came to the conclusion that if circulation in the Atlantic Ocean were to cease, the resultant cooling would be so minor that global warming would outpace it within decades.

That is not to say that melting events are always insignificant. Water has the highest heat capacity of all known liquid substances, meaning that the oceans play a primary role in the storing and transporting of thermal energy around the planet. If cold, upwelling currents in one part of the world cease because they are capped by warm, buoyant surface waters, then it would not take long for global temperatures to rise. This is indeed precisely what happens after each El Niño event, when higher temperatures in the eastern Pacific put a temporary stop to the cold, upwelling Humboldt current from Antarctica. There have been about twenty El Niño events over the past century, with each one causing a noticeable spike in global mean temperatures.[5]

The phenomenon of ocean stratification, whereby water bodies can be physically separated from one another, is known to occur in all of the world's oceans, with each flowing layer defined by its unique combination of temperature and salinity. Strictly speaking, persistent stratification is only a deep ocean phenomenon because storms can whip up the shallows, rapidly destroying any vertical density contrasts. The "storm wave base," the depth to which surface turbulence no longer has any effect, can reach down as far as a hundred meters. Beneath the storm wave base, seawater typically gets colder and therefore denser, particularly beneath about 1,000 meters, where the "pycnocline" effectively separates the upper layers from the deep ocean, at least where there is no upwelling or downwelling and on time scales of tens to hundreds of years. Melting events caused, for example, by ice sheets breaking off from Antarctica or Greenland are unlikely to release sufficient quantities of water quickly enough to stop vertical ocean mixing for long. This is mostly because the world's great polar ice caps have largely remained in place since the recent glaciations began, with most melting occurring at intermediate latitudes. But what if deglaciation was global, irreversible, and rapid? What would happen then? Could the oceans ever be physically stratified for long periods on a global scale?

Although outrageous, this may be precisely what happened at the end of the Cryogenian ice age, as it would most parsimoniously explain so many otherwise inexplicable features of cap dolostones: their global occurrence, their knife-sharp contacts on top of fresh-looking periglacial landscapes across the world, and even their strange mineral makeup. Once meltwater can occupy the surface ocean down to several hundred meters, the time scale for mixing becomes much longer than even Hollywood scriptwriters would dare to imagine, effectively sealing off the surface environment from the deep oceans for tens of thousands of years. To envisage such an incredible circumstance, we need to abandon the geologist's traditional mantra, that "the present is the key to the past," and adopt instead the well-known adage from L. P. Hartley's novel of 1953, *The Go-Between:* "The past is a foreign country: they do things differently there."

Earth's past really is a foreign country. In the immediate aftermath of runaway deglaciation, the deep ocean would inherit strange features from the ancient ocean of Snowball Earth. With so much water locked up in ice, the syn-glacial ocean would have been highly saline, freezing cold, and extremely dense, properties that would be retained in the post-glacial deep ocean. Effectively decoupled from the warming surface environment, the deep ocean would grow increasingly estranged from its surface counterpart, eventually losing any remnant of the free oxygen it might once have held. With time, all oxidant would be used up by reaction with reduced volcanic gases at mid-ocean ridges and sinking organic matter. As soon as all oxygen was gone, sulfate would become the next victim as the deep oceans of the world gradually turned into an anaerobic microbial soup. Anaerobic bacterial ecosystems, reminders of a much earlier age but today relegated to anoxic swamps, stomachs, and sediments, would have come out of hiding, turning the deep ocean into a stinking sulfide stew, like a global Black Sea. These were perfect conditions for barium concentrations to build up in, entirely separate from the overlying sulfate-rich meltwater plume. When the two water bodies did finally mix, presumably in zones of deep, upwelling anoxic waters, barite was the inevitable result. This scenario also explains the strange association between barite and the pinkish dolomite, as the upwelling, anoxic waters would also have been laden with ferrous iron (Fe II) oxidized to reddish Fe (III) in the surface waters. Together, they provide strong circumstantial evidence of this extremely alien world on earth.[6]

A surprising but inevitable consequence of this wild tale is that the carbonate minerals in cap dolostones the world over, no matter whether deposited hundreds of meters below the edge of the continental shelf or across the much shallower coastal platform, were all deposited beneath a freshwater lens. Sounds crazy, but it is perhaps not as ridiculous as all that. Carbonate minerals are not keen to precipitate even in today's marine environment, even though seawater is highly oversaturated with respect to all major marine carbonate minerals (calcite, aragonite, and dolomite). This is because their precipitation is inhibited by many factors, especially the various salts dissolved in seawater. In most cases, calcite and aragonite precipitate only when coaxed into doing so by organisms that build their shells from these minerals. Before shells evolved, life could only mediate the process through the activities of microbes, which might incidentally cause alkalinity to rise locally in either the water column or the sediment. Indeed, this is the only way dolomite can form at normal surface temperatures and pressures today. In the absence of such help, only abiotic precipitation can take place, but for that to happen seawater needs to be hugely enriched in calcium, magnesium, and carbonate ions.

Freshwater avoids many of these problems. Each summer in mountain lakes around the world, microscopic plants bloom, fertilized by the nutrients brought in after the last of the snow has melted, exposing glacial flour to erosion by wind and rain. The tiny picoplankton take in carbon dioxide, as do all plants, for their photosynthesis, and in so doing shift the pH of surface waters by enough that calcium carbonate rains down like snow onto the lake floor below. Each bloom turns the lake waters pale in so-called whiting events, which help to date the lake's annual sedimentary layers in the same way we might count tree rings. Although carbonate precipitation is normally associated with hot climes, just think coral reefs, whiting events do not always occur in high summer. For example, in Hagelseewli, a high mountain lake in Switzerland, whitings usually occur at water depths of six to nine meters and temperatures of only 4 degrees Celsius, a good demonstration that carbonate precipitation is almost always a biologically mediated, rather than entirely inorganic, process.

Lakes such as these are good analogues for the post-glacial, meltwater-dominated, surface ocean "plume world." The low salinity of freshwater greatly reduces mineral solubility, meaning that carbonate alkalinity could not build up to concentrations high enough to prevent the pH changes caused by photosynthesis. The freshwater plume world would have been a place where bacterial blooms triggered whiting events the world over. During the early stages of transgression, the seafloor would have been covered by microbial mats, whose surface layers rapidly calcified in the warming ocean margins, building stromatolite towers as they raced to keep up with the rapidly rising sea levels. Once submerged below the photic zone, cap dolostones would have found themselves in deeper waters far from the coast, where the strengthening physical stratification prevented oxygen from reaching the seafloor. Microbial communities changed into anaerobic ecosystems, the like of which had not been seen for billions of years. Geochemical studies have discovered the isotopic fingerprints of both methanogenesis by anaerobic archaea and sulfate reduction by anaerobic bacteria in cap dolostones. These are perfect conditions for dolomite precipitation, which today is prevented by the presence of high concentrations of sulfate ions in seawater. Indeed, methanogenic archaea are known to assist dolomite precipitation, providing the only known way to do this under the unique combination of conditions that probably would have existed in the plume: that is, low temperatures and neutral to slightly acidic pH values. In many areas, the pressure release caused by isostatic readjustment led to tension cracks, which cemented shut even as they opened. The same pressure release caused volcanic eruptions, earthquakes, and submarine landslides, evidence for which can be found throughout northwest Africa and Brazil, for example. However, the cap dolostone in Senegal is even more special in that in some places it consists entirely of volcanic debris, which slid into deeper realms where the volcanic ash and shards turned into dolomite.[7]

After many years of head-scratching by geologists like Max Deynoux and Bob Dalgarno, scientists are slowly reaching a consensus opinion that cap dolostones are an otherworldly but predictable consequence of runaway deglaciation, the scale of which has never been seen before or since. The picture that has been built up through detailed geological studies points to an exceptionally large and rapid injection of meltwater into the world's oceans at the end of the Cryogenian ice age. Numerous geochemical and isotopic studies have come together recently in support of this idea, and we will learn about some of these in a later chapter. A vindication for Brian Harland, perhaps, but this is far from the end of the debate about his Great Infra-Cambrian Glaciation. Although their presence may be predictable, the enormous scale of cap dolostones is not, and their size implies extraordinarily high rates of carbonate precipitation. The amount of carbonate locked up in cap dolostones is estimated at more than one million gigatonnes, almost an order of magnitude more than all the carbonate in today's oceans, atmosphere, and soils combined. The typical features of cap dolostones, the towering but skinny stromatolites, the early cemented tepees and sheet cracks, demonstrate unusually high levels of calcium carbonate oversaturation. How so much carbonate alkalinity built up so quickly in the aftermath of glaciation has been the subject of considerable controversy, as we shall soon see, but it is important that we do not leave cap dolostones without addressing what I regard to be their crowning glory.

The belated consensus that cap dolostones formed during global deglaciation implies that they represent wonderfully precise time markers, prized treasures in stratigraphy. The basal Ediacaran "golden spike," or GSSP, is among the most highly resolved time markers of the entire geological time scale. Because cap dolostones formed above glacially influenced sediments in all cases, their presence suggests that few places on earth, if any, had escaped ice sheets during that time. This raises the question, just how extensive was the Cryogenian glaciation? How much ice was locked up in ice caps on the continents, and for how long? What were the consequences of glaciation's catastrophic aftermath for life on earth? The answers to these questions shed light on events far beyond climate change. The stage is set for the Snowball Earth hypothesis.

4

Frozen Greenhouse

Some 717 million years ago, in parts of the world, glaciers scoured directly into tropical carbonate reefs, implying either that continents had moved rapidly from the equator to the poles or that temperatures had plummeted even at low latitudes. After geophysical data confirmed the once crazy notion, it was but a short step to postulating a global glaciation. The Snowball Earth hypothesis proposes that global ice cover allowed carbon dioxide to reach extraordinary levels before the greenhouse effect could finally overcome the high albedo of a completely frozen planet.

Many geology students the world over undergo a rite of passage before becoming fully fledged geologists. The requisite task, to make a geological map from scratch, has remained almost unchanged since the 1950s and is still an accreditation requirement of the Geological Society of London. It was thus in a long and honored tradition that I was sent forth by London's Royal

School of Mines in 1990 to do my own formative apprenticeship and make a map, which was to cover a small area, about
five kilometers by five kilometers, of the island of Islay off
Scotland's west coast. During my six weeks of traipsing windswept hills and coast, I was the sole occupant of a disused caravan in a field, with only ticks, midges, horseflies, and herds
of sheep for company. On my return, strangely bronzed by the
outdoor life despite the incessant drizzle, my admiration for
field geologists was raised its first notch. The admirably detailed
work of one person, Tony Spencer, caught my eye in particular. Two decades earlier he had researched the same area and
found abundant evidence for late Precambrian glaciation in
Scotland.

Following the publication in 1964 of Brian Harland's case
for a Great Infra-Cambrian Glaciation, ice ages in the tropics
had become a hot topic, and evidence was soon pouring in
from all around the world. Bob Dalgarno and his colleagues
at the Australian Geological Survey had just completed their
detailed mapping of the Flinders Ranges in South Australia,
which confirmed the claims of the late, great Douglas Mawson.
Max Deynoux and a group of young French geologists were
also adding to the case for glaciation in the Sahara. However,
despite some early attempts by none other than Brian Harland
himself, there was as yet no firm geophysical evidence in support of his proposal for truly global glaciation, or, in other
words, for low-latitude glaciers at sea level. I knew that Max
had worked just as hard building evidence for the late Ordovician glaciation, which had also affected the Sahara. However,
it was well established by then that northern Africa had been at
the South Pole during the Ordovician Period, so no surprises
there. Many geologists believed that continental drift would explain the Cryogenian ice age, too, if it were not for one ugly fact
that always stood in the way: their close association with car-

bonate rocks, which are more commonly associated with high temperatures and low latitudes. Ian Fairchild, now a professor emeritus at Birmingham University, summed up this apparent contradiction in 1993 with the juicy phrase "balmy shores and icy wastes."[1]

Key to those early arguments was Tony Spencer's comprehensive study of my undergraduate mapping area that has since become a classic of paleo-glaciology. Originally published in 1971 as a marvelously unpretentious memoir of the Geological Society of London, it helped to launch the oil industry career of this engaging and energetic Yorkshireman. Half a century after beginning his forensic interrogation of the glaciogenic diamictites of Scotland, Tony can still be found striding the hills and glens of Islay and the nearby Garvellach Isles, now equipped with a walking stick, bright white crash helmet, and an enthusiasm in retirement that is as contagious as it is obsessive. In recent years, Tony has brought together a growing entourage of students and professionals alike, including Ian Fairchild and myself, to take another long, hard look at these wonderfully informative rock successions.[2]

The rocks in question belong to a group called the Dalradian Supergroup, which is an enormously thick pile of sediments that can be found stretching across Scotland and Ireland in a band from southwest to northeast. The unusual thickness of this succession is due to its location on the rapidly subsiding margin of the ancient Iapetus Ocean. Rapid subsidence is key to preserving a relatively complete record of geological events, so such places are highly prized. Although complete successions might be best preserved in deeper marine environments, it is rare for these to survive destruction through metamorphism or subduction. The thick Dalradian succession therefore offers one of the only possibilities in the world for preserving a transitional record of the onset of glaciation. In both northwest

Africa and Australia, the descent into glaciation is marked by erosional scours that have cut into much older rocks below. My caravan, rented from a local family, was situated in the middle of the field area on pre-glacial rocks, in full view of rocky crags of the glacially derived Port Askaig Formation to the north. This was the location of the very first report, way back in 1871, of Precambrian glaciation, which sparked the controversy that is still hotly debated today.

Western Scotland is about as different from the Sahara Desert or the Australian outback as is possible to imagine. However, back in the pre-Cryogenian Neoproterozoic Era, Scotland looked rather different from today. Most of the rocks Tony and Ian have studied beneath the diamictites of the Port Askaig Formation are limestones. Close to my caravan, a quick check with the trusty hand lens revealed that the limestone nearby comprised tiny balls made up of concentric rings of the mineral calcite (calcium carbonate). Ooids like these have formed throughout most of earth history but are found today only along balmy, tropical shores like the Bahamas, where they grow by accreting layer upon layer of calcium carbonate while being tossed about like sand grains in the tidal currents. Although I encouraged you to abandon thoughts of uniformitarianism in the previous chapter, it is hard to escape the conclusion that, in Scotland, glaciers rode roughshod over tropical shoals, normally associated with low latitudes. A superb example of the sheer power of glacial scouring is "The Bubble" on Garbh Eileach, also known as Holy Isle, off the western coast of Scotland (Figure 5). This improbable juxtaposition of hot and cold environments was one of the main objections that led Schermerhorn to ridicule the notion that there had ever been a late Precambrian glaciation.

Tony Spencer has located the contact between the carbonate platform and the overlying diamictite at the northern-

Figure 5. "The Bubble" on Holy Isle. The "bubble" (contorted, white rock in photo) is part of the Great Breccia (Port Askaig Formation) on Holy Isle (Garbh Eileach) off western Scotland. This huge chunk of rock about 100 meters across was scoured off the underlying limestone reef, before being squeezed and contorted to form a "recumbent" fold under tremendous pressure from the overriding ice sheet during the Cryogenian Period. (Photo by author, 2013)

most tip of the island of Garbh Eileach. On a calm day, local boatmen will take you to the island for a fee, just as they have always done, but you'll have to jump to make land as the only jetty is on a different part of this uninhabited isle. The neighboring island of Eileach an Naoimh, also composed of diamictite, was once home to Irish monks whose chapels and hermitages, the oldest ecclesiastical buildings in all of Scotland, can still be seen standing proudly against the ravages of the weather. Together these islands make up the Garvellachs, where Dalradian rocks have curiously escaped the worst of the tectonic deformation caused by the closing of the Iapetus Ocean that eventually led England to smash into Scotland. This could

Figure 6. Anthony Spencer, standing on outcrops of the Keills
Member of the pre-glacial Lossit Limestone Formation, south of
Beannan Buidhe on Islay, looking eastward at the rounded hills, or
"Paps," of Jura. The photo was taken after the discovery of a form of
characteristically pre-Cryogenian calcite (molar tooth structure) at
this locality. (Photo by Ian Fairchild)

be one of the most likely places on earth to find an unbroken
record of our planet's descent into the Cryogenian deep freeze.

With Tony as your expert guide, you can almost feel the
chill set in as you jump off the boat to land on layers of lime
mud showing the unmistakable signs of microbial mats and
the occasional stromatolite domes (Figure 6). Diving amid these
tropical reefs with their predictable riot of color would have
been a joy, reminiscent of the much older stromatolite reefs
of Mauritania, but conditions were just about to change. The
immediately overlying carbonate rocks show clear signs that
the coastal environment was getting shallower over time. It is

Figure 7. Ice wedges on Garbh Eileach. Periglacial, polygonal,
sand-filled cracks formed by freezing and thawing during the
Cryogenian glaciation, Garbh Eileach, western Scotland. (Photo
by author, 2013, with glaciologist Marie Busfield on the left, and
geologist Galen Halverson on the right)

tempting to think that these were the first signs that ice was
forming at higher latitudes, causing sea levels to fall before the
onset of full-blown glaciation. The expansion of ice cover is
evidenced by hundreds of meters of diamictite, revealing cycle
upon cycle of frost-heaved patterned ground and mass flows
(Figure 7). The Garbh Eileach succession records such an abrupt
climatic shift that it leaves no time for continental drift to move
these tropical shoals to the poles. In this area, at least, the Cryogenian Period arrived with a bang rather than a whimper.

For what it's worth, I wholeheartedly agree with the Dalradian evangelists Tony, Ian, and others who see little or no
cryptic break in this succession, which, taken at face value, implies that the onset of Cryogenian glaciation was disconcert-
ingly abrupt. In recent years, Ian has sought to prove this with

a small touch of genius, by analyzing the cobbles and pebbles deposited from the overriding ice sheets. In his grand memoir, Tony had described in great detail how the composition of outsized clasts changed over time, being mostly dolomitic at the base and mostly granitic at the top of the diamictite. These bright red granite boulders are the same ones that James Thomson, whom we met in chapter 1, remarked upon in 1871, but few people have taken much notice of the dreary dolomitic clasts, presumably because they are of much more local origin, having been "plucked" directly from the underlying rocks. Ian predicted that the isotopic composition of the clasts through the lowermost diamictite (working upward from its base) would present a mirror image of the carbonate succession beneath. This is because the first deposited boulders ought to represent the first rocks to be eroded, which should also have been the highest. And that is precisely what he found. There is no sign of any missing strata, in the form of unexpected clast types or isotopic compositions, confirming a negligibly small time gap at Garbh Eileach.[3]

Recently, other isotopic evidence has confirmed the transitional nature of the Dalradian succession. We will take a more detailed look at such evidence in a later chapter, but for now it suffices to note that the Dalradian rocks are far from the only example of Cryogenian glaciers riding over carbonate platforms. Some fellow geologists consider the base of the Cryogenian System in Greenland or Spitsbergen to be even more convincing examples of tropical glaciation, and some prefer the equally expanded sections of the Canadian Rockies. Others dispute that any of these can truly be complete records, and argue that we must look to deeper marine environments, such as can be found in China, to view a genuinely transitional section, but there are precious few of these and none that contain carbonate rocks. To seal the case for global glaciation, once and for all, we need to turn to geophysics. Building on the considerable circumstantial evidence of tropical ice cover, it was the study of the earth's past magnetic field that eventually led to a resurgence of interest in the Cryogenian ice age, and that in turn led to a convincing explanation in the "Snowball Earth hypothesis" for why glaciation at low latitudes was so catastrophic.

Talk to many Australians, especially those from the tropical north, and some will tell you how they swam with freshwater crocodiles in their youth and had grown accustomed to all manner of nasties from snakes and spiders of the outback to the toxic cone shells, lionfish, and box jellyfish of the Great Barrier Reef. One of my abiding memories from a trip in 2007 to the Kimberley Ranges of northwestern Australia was of doing just that, washing and swimming alongside freshwater crocodiles, and in the process becoming remarkably nonchalant about it, to the consternation of my wife back home. The general wisdom is that although freshies might nibble, that's an acceptable inconvenience because the much nastier salties wouldn't let you out alive. Our aboriginal guides regaled us with gory crocodile tales, and we left that place with a greater appreciation of nature at its most unrestrained. My friend and guide on that trip, Maree Corkeron, had spent many months working in the Kimberleys and is an expert on the glaciations and cap dolostone there. While working on her doctorate, she even took her husband and two young daughters with her on her long outback field trips.

The trip was organized under the auspices of another of those international UNESCO projects, IGCP 512, this one dedicated to the late Precambrian ice ages. As well as Australians, there were several other nationalities on display, including some old friends from our previous outback experience just a few years before. Joining us were two geophysicists, specialists in the earth's changing magnetic field, Joe Kirschvink and Mark

McWilliams. Both men had completed their respective PhDs on the ancient magnetic field in Australia—clearly a hot topic in the late 1970s—and so were old rivals. I think it fair to say that, despite coining the term "Snowball Earth," Joe started off as a skeptic of global glaciation, and not without cause. Since the heady days of the late 1950s, glacial deposits had become a favorite target for paleomagnetic studies. Early on, it became established that iron-rich grains in sediments orient themselves to the earth's magnetic field. Once turned into rock, the preferred orientations and information about the paleomagnetic field are locked in for all time and so can be used to reconstruct the latitude at which those sediments were deposited. However, it was not so easy to determine whether the latitudes thus obtained had truly remained unaltered, or whether over time they had been reset by all manner of disturbances.

Because it was believed to have formed at low latitudes, the Port Askaig Formation in Scotland was a frequent target of study throughout the 1970s, yielding the consistent result that its deposition had taken place within ten degrees of the equator. Yet doubt remained even among global glaciation's most earnest supporters, for there was the nagging suspicion that the heat generated by the much later collision between Scotland and England had removed all traces of the primary magnetic field. We now know that such "overprinting" had indeed happened, and that all previously published magnetic data from Dalradian rocks relate to Scotland's position during the Ordovician and not the Cryogenian Period. The skeptical nature of scientists led them to establish ever more rigorous tests, which needed to be passed in order for paleolatitudes to be trusted. The first robust tests of low-latitude glaciation came in quick succession in 1986 and 1987, by two teams working on the same glacial deposit at Pichi Richi in South Australia. Crucially, both studies reported equatorial latitudes for glaciation,

but Joe Kirschvink's team went one step further. In an ingenious experiment, they showed that the latitudinal information
they obtained had been disturbed by a landslide, or slump, before it had turned to rock, a clear sign that the paleomagnetic
properties were original features of the deposited sediment.
Such "fold tests" now form part of the standard repertoire of the
scientists we geologists fondly refer to as "paleomagicians."[4]

Having had his road to Damascus moment, Joe couldn't
be stopped, and in 1992 he bet his reputation firmly on global
glaciation, by framing in two pages the bare bones of what
came to be called the "Snowball Earth hypothesis." After testing the evidence, he could claim with some confidence that
glaciers had indeed reached sea level even near the equator; it
was a small step from there to suggesting that the entire globe
had been covered in ice. Many years earlier, it had been shown
that ice sheets would rapidly expand to cover the equatorial
regions should they ever encroach near the earth's midriff. This
is because the brightness of ice tends to reflect solar energy away
from the earth, causing a runaway positive feedback once ice
cover reaches the low-latitude areas that receive most of the
sun's rays. Once the entire planet was ice-covered, it would
have taken a very long time to recover, Joe surmised, because
the effect of the increased albedo could only be counteracted
by a gradual buildup of greenhouse gas over millions of years.
With ice covering both land and sea, chemical weathering and
photosynthesis, the world's major carbon sinks, would have
been suppressed, allowing carbon dioxide levels to increase
far beyond what would normally be possible. Joe Kirschvink's
Snowball Earth hypothesis makes the counterintuitive prediction that the Cryogenian Period was a time of extremely high
levels of greenhouse gases in the earth's atmosphere. Both pro-
vocative and speculative, that central tenet of the hypothesis is
now widely accepted.[5]

After my mapping in Scotland, I moved to Switzerland to start my own PhD. That same year, following Joe's lead, there was growing excitement among geologists about the possibility that our planet had once been entirely covered by ice. Geophysicists took up their positions for and against over the following few years, but, in 1997, a paleomagnetic study of the Mackenzie Mountains in northwest Canada was published that finally clinched the argument once and for all. The section that was studied, much like the one at Garbh Eileach, evidenced glaciers scraping the top off of a carbonate platform. The study proved that both carbonates and diamictites were deposited within just a few degrees of the equator. The time was ripe to bring these findings to the attention of the wider scientific community. Enter Paul Hoffman, one of the world's great living geologists and human dynamo. As well as numerous other claims to fame, Paul seems to have been the first person to make the connection between Kirschvink's Snowball Earth hypothesis and the weird "cap carbonates" that seemed always to appear once glaciation had ended. He had been investigating Cryogenian glaciation in Namibia for several years, and in 1998 his team at Harvard University finally published the idea that the carbonate in cap dolostones and limestones derived ultimately from the hyper-greenhouse atmosphere inherited from Snowball Earth's near shutdown of the global hydrological cycle. Fierce Snowball fights dominated the field for the next decade or so, leading to an increasingly nuanced framing of the hypothesis that we will come to in due course.[6]

The Snowball Earth hypothesis finally explained why the deglaciation had been so abrupt and seemingly without pause. Once ice began to melt at mid-latitudes, it did so at an ever quickening pace due to the combined effects of decreased albedo and a renewed water cycle, accelerated by the anomalously large amount of carbon dioxide, which strengthened the greenhouse effect and made a return to glaciation impossible. As the surface ocean warmed, carbon dioxide degassed from the oceans into the atmosphere, pushing temperatures even higher. Both degassing and warming encourage the precipitation of carbonate minerals, injecting yet more carbon dioxide back into the atmosphere. Carbon dioxide could not fall back to more normal levels again for some time after runaway deglaciation due to an array of positive feedbacks like these that would likely have taken hundreds of thousands of years to play out. If we want to understand why it takes so long for surplus carbon dioxide to be removed, we need to take a short detour into the workings of the long-term, or geological, carbon cycle. Much later in the book, we will return to this topic because long-term carbon cycling affects not just climate, but also the oxygen that our earliest animal ancestors needed to breathe.

The long-term carbon cycle was first outlined two centuries ago during a period of time that hindsight has termed the Enlightenment, but which coincided with revolution, counter-revolution, and outright war. One of the early casualties was the man who, having named all of the most important life-giving elements, was then recklessly sentenced to death, guillotined, and pardoned; tragically, in that order. One of the last of Antoine Lavoisier's many achievements was to show that Joseph Black's mysterious "fixed air," which Carl Scheele would later call "aerial acid," was actually made of two of his newly discovered elements, carbon and oxygen. In 1789, in his grand treatise, Lavoisier was thus able to describe carbon as an "oxidisable and acidifiable non-metallic element." Joseph Lagrange, writing after the death of Lavoisier, mourned the loss to science, writing that "it took them only an instant to cut off his head, but France may not produce another such head in a century."

Although the sentiment can only be applauded, it was more difficult than Lagrange imagined to halt the Enlightenment juggernaut. In the fifty years following Lavoisier's sorry demise, many great French scientists showed themselves more than capable of taking up the mantle laid down by the great man. Joseph Fourier predicted in 1827 the warming effect that carbon dioxide has on our atmosphere, and already by 1847 Jacques-Joseph Ebelmen had described a complete account of the global carbon cycle, employing the same chemical equations that we use today. The names of all these men, along with other pioneers in our understanding of the carbon cycle, such as Brongniart, Boussingault, and Dumas, appear on the Eiffel Tower. It seems incredible, looking back, that the intricate workings of the long-term carbon cycle, including its governing of atmospheric composition, and therefore climate, on million-year time scales, were discovered even before Darwin had published his *Origin of Species*. Sadly, as we shall see, the outline of the long-term carbon cycle had to be rediscovered a century or more later.[7]

These days, we are all aware that carbon dioxide emissions cause global warming, but there is nothing inherently sinister about the greenhouse effect. Without the warming influence of atmospheric carbon dioxide, the earth's mean surface temperature would plunge more than 30 degrees Celsius. Nevertheless, by virtue of there being only a relatively small amount of carbon dioxide in the atmosphere, now just over 400 parts per million (ppm) by weight, the greenhouse effect is a fickle friend indeed. Short-term changes in carbon dioxide levels, for whatever reason, are now prime suspects in global catastrophes, such as hyperthermal, glacial, extinction, and asphyxiation events throughout earth history. No geological text can therefore afford to ignore the carbon cycle, but its nature differs greatly in terms of magnitude and mechanism depending on the temporal and spatial scales over which it is viewed. In other words, it is very important not to conflate the very different processes that operate on time scales ranging through ten orders of magnitude.

For me, the most awe-inspiring feature of the familiar exponential rise in atmospheric CO_2 since the Industrial Revolution, and its tremendous steepening since the 1960s, is the way annual cycles of planetary breathing, particularly in the land-dominated northern hemisphere, can so easily be distinguished. The deep photosynthetic inhalation of carbon dioxide during the northern hemisphere's spring matches an almost equal release during the fall months. We call this latter surge "respiration" because it mimics to some extent what our lungs do, but we might also call it decay. This is what geologists refer to as the short-term carbon cycle. All things being equal, the annual ups and downs of the short-term carbon cycle should balance out, but humanity's fossil fuel burning means that each spring's low point in the cycle is now just a little bit higher than the year before. Ignoring the anthropogenic influence, almost a tenth of the atmosphere's carbon dioxide inventory passes through the biosphere each and every year, but carbon dioxide is also cycled around the planet on much longer time scales. Although there are several different overlapping cycles, we tend to lump them all together as the "long-term carbon cycle," which we think takes 10,000 times longer to reach equilibrium.

I like the term "fossil fuel" because its meaning is so self-evident. Coal, oil, and gas are quite simply the fossilized remnants of organisms that were once living. Most sediments, whether marine or terrestrial, retain some organic material, commonly with as much as a few percent or more of carbon by weight. In many cases, the organic carbon in rocks can be extracted to reveal exquisitely preserved fossil spores and plankton or even the molecular fingerprints of chlorophyll and other organic compounds diagnostic of life. In order for such fossil carbon to exist, the seemingly perfect balancing act of the annual short-term carbon cycle cannot be quite so perfect after all. In other words, it has a leak, which although small, is significant over longer time scales. If left unchecked, this leak would eventually drain all the carbon dioxide from the air. I am not sure whether we would freeze or starve first, but as this has never once happened to our ancestors, the leak must be compensated for by another process, which must continually add carbon dioxide to the atmosphere. Part of the answer to this quandary lies in the oxidative weathering of fossil organic matter in soils, which is the flip side to organic burial. But this "organic long-term carbon cycle" is only part of the story. It was Jean-Baptiste Boussingault, a contemporary of Ebelmen, who first surmised that volcanism must provide a second source. Volcanic outgassing of carbon dioxide released from the mantle causes acid rain, which attacks the rocks on the surface of the earth, dissolving mountains grain by grain. This chemical weathering of rocks was well known by the middle of the nineteenth century through discoveries made largely by German chemists such as Gustav Bischof, who showed that a simple mixture of carbon dioxide and water (carbonic acid) could weather even tough granite to soapy clay given enough time.[8]

Building on all these foundations, Jacques-Joseph Ebelmen put the long-term carbon cycle together. Ebelmen realized that all net sources of carbon dioxide (oxidative weathering and volcanic outgassing) must equal the net sinks over geological time scales, and he correctly identified those long-term sinks as the deposition of organic and carbonate carbon on the seafloor. He surmised, quite rightly, that even a tiny imbalance between source and sink would lead to such massive environmental changes as to be implausible. As Bischof argued, it is chemical weathering that links source to sink because the dissolution of silicate rocks releases calcium ions, carbonate alkalinity, and nutrient phosphate into the marine environment. The first two of these combine to form calcium carbonate, while nutrients are what drives organic production, and therefore organic carbon burial. Not all deposited carbonate is a net sink for carbon dioxide on long time scales. The weathering of carbonate rocks does lead to carbonate deposition, but it does not have the same effect as the weathering of silicate rocks because the acidity used in limestone dissolution is released again when carbonate minerals precipitate. Indeed, the coupled precipitation and dissolution of carbonate minerals on land and in the sea represents a buffer against carbon cycle perturbations that acts on medium time scales of only a thousand years or so. By contrast, when we compare the net carbon flux through the ocean-atmosphere system with the amount of carbon in surface reservoirs, we can estimate that the silicate weathering–carbonate deposition cycle achieves equilibrium on much longer time scales of around a hundred thousand years (Figure 8). This means that excess greenhouse emissions today will lead to anomalously high atmospheric carbon dioxide levels for a very long time to come.

The invention of the steam engine taught the leaders of the Enlightenment movement a valuable lesson, which is that natural cycles also have to be perfectly balanced, at least on long time scales. In other words, through checks and balances, perfect economy can arise naturally out of a complex system through negative feedbacks. Adam Smith described this phenomenon in 1776 as the "invisible hand" in his influential book *The Wealth of Nations*. This is why Ebelmen did not ask himself whether the carbon cycle was balanced, but only how it was balanced. If he had lived beyond the tragically young age of thirty-seven, he might well have discovered the full story of how carbon cycle feedbacks regulate climate. The answer was eventually provided much later, however, by a seminal and, in the curious manner of such things, surprisingly short paper

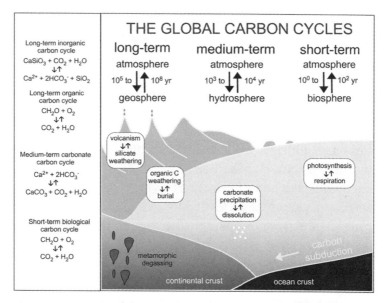

Figure 8. The global cycles that govern the flow of carbon
dioxide through the earth's ocean-atmosphere system interact on
various time scales. For example, the long-term sinks for atmospheric carbon dioxide are (1) burial of organic carbon, originated
from photosynthesis, and (2) calcium carbonate, derived from
weathering. Buried carbon returns only millions of years later
following subduction and metamorphic outgassing, or tectonic
uplift and oxidation.

entitled "A Negative Feedback Mechanism for the Long-term
Stabilization of Earth's Surface Temperature." The mechanism
published in 1981 by three American geoscientists, James Walker,
Peter Hays, and Jim Kasting, has since become established as
the working paradigm for understanding how the long-term
carbon cycle regulates global climate. In essence, they took the
geological carbon cycle—it had since been rediscovered by various people, including the Nobel Prize laureate Harold Urey—
and added to it a temperature sensitivity related to the green-

house effect. In the way of all great ideas, it seems deceptively simple in hindsight.

The crux of this negative feedback lies in the addition of a temperature control to Ebelmen's long-term carbon cycle, linking global chemical weathering rates to the amount of carbon dioxide in the atmosphere. Let us consider the case of a decrease in the greenhouse effect. If this were to occur, then global temperatures would start to fall. The resultant drop in chemical weathering rates would then allow carbon dioxide levels to rise again, thus thwarting perpetual deep freeze. Conversely, if new volcanic provinces were to erupt, spewing out millions of cubic kilometers of lava, the enhanced greenhouse effect from all the outgassed carbon dioxide would overheat the earth. Carbon dioxide solubility would be lowered by the warmer conditions, thus driving a positive feedback, whereby higher temperatures degas ocean carbon dioxide into the atmosphere, causing even higher temperatures and wetter skies. However, balance would eventually return because chemical weathering is greatly favored under a warm, humid, and more acidic atmosphere. Under such greenhouse conditions, the extra carbon dioxide in the air would be titrated away, to be deposited as carbonate on the seafloor. Only through tectonic cooking (called metamorphic decarbonation, potentially through subduction) could the stored carbon dioxide ever be released again, thus closing the cycle tens to hundreds of millions of years after the initial eruptions (Figure 9).

For this natural thermostat to work, there needs to be an active water or "hydrological" cycle, which can accelerate both physical and chemical weathering. On a perpetually glaciated planet, however, it is hard to imagine how weathering could keep up with outgassing, and so, as Kirschvink proposed, carbon dioxide levels in the atmosphere would have kept building up. After deglaciation, the water cycle would spring back

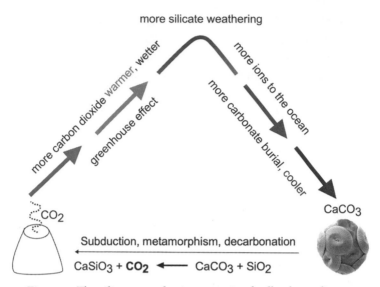

more silicate weathering

more carbon dioxide warmer, wetter

greenhouse effect

more ions to the ocean

more carbonate burial, cooler

CO₂

CaCO₃

Subduction, metamorphism, decarbonation

$CaSiO_3 + CO_2 \longleftarrow CaCO_3 + SiO_2$

Figure 9. The silicate weathering negative feedback on climate. Climate perturbations are regulated by changes to chemical weathering, riverine flux, and carbonate deposition within a hundred thousand years.

into action, not just to normal levels, but supercharged by the ultragreenhouse. The high acidity of rainwater and spiraling temperatures would have ensured an unprecedented surge in chemical weathering rates, as the finely ground glacial flour and cleanly etched glacial surfaces became exposed to attack. Given enough time, increased weathering would lead to more carbonate deposition, but cap dolostones are not that particular smoking gun. As we saw earlier, cap dolostones were formed during deglaciation, on time scales too rapid to be dominated by the titration of this excess carbon dioxide. Carbonate precipitation seems to have been so rapid that even cap dolostone deposition would have been a net source, rather than sink, of carbon dioxide during those earliest stages of the glacial aftermath, turning the surficial plume world even more acidic. Geochemical studies, which suggest that pH dropped by between one and two whole pH units during cap dolostone deposition, lend credence to this notion.[9]

This latter point is possibly not immediately evident. Surely, carbonate deposition removes carbon from the ocean-atmosphere system, and so would reduce levels of carbon dioxide. Not so. As Ebelmen correctly divined so long ago, carbonate deposition is only a net sink for carbon dioxide when linked to silicate weathering. Because surface oceans are saturated with respect to carbonate minerals—that's why reefs have been perpetual features of the world's oceans for billions of years—almost every calcium ion that is released during weathering will eventually couple with two bicarbonate ions to form a carbonate mineral, thereby releasing a molecule of carbon dioxide.

If carbonate precipitation occurs in equilibrium with carbonate weathering, then the carbon dioxide released during precipitation will approximately equal that used up in chemical weathering. However, if excess carbonate precipitation occurs on time scales shorter than 10,000 years, carbon dioxide release will outpace weathering, at least for a time. In short, the very existence of rapidly precipitated cap carbonates is consistent with unstoppable deglaciation. If greenhouse gas levels were high before deglaciation, they would have been even higher afterward, overheating the brackish meltwater plume, and unceremoniously tearing the planet out of the deep freeze and into the fire.

Even though the oceans are tremendously deep, heat exchange between the surface plume world and the colder lower ocean meant that it was only a matter of time before the physical stratification of the oceans broke down. Cooler, saline waters would have upwelled over the shelf edges, finally degassing their load of carbon dioxide into the atmosphere as they warmed.

An ocean in equilibrium with a carbon-dioxide-charged atmosphere would inevitably become rich not only in carbon
dioxide but also in bicarbonate. The combination of degassing
and carbonate alkalinity ensured deposition of a new suite of
carbonate rocks; this time over the edges of the now drowned
continental shelves.

The signs of this upwelling have been puzzled over for
thirty years or more. Strange crystal fans of what was once aragonite (a metastable calcium carbonate mineral) sit on top of
cap dolostones in many parts of the world and in some places
are intimately related to the seafloor barite mentioned in the
previous chapter. Such crystal fans have been reported from
post-glacial sections in Brazil, Canada, the United States, and
Namibia, but, unlike cap dolostones, do not occur worldwide.
Aragonite fans are highly unusual features of the ancient seafloor. Their clearly abiotic origin indicates extremely high levels of oversaturation. The only other similar occurrences have
been reported from much earlier times in earth history and
have been interpreted as reflecting anoxic conditions and the
inhibiting properties of high levels of dissolved ferrous iron.
The location of these fans at precisely the places and times
where we already envisaged upwelling of anoxic waters from
the strange barite and iron-rich dolomite mineral associations
mentioned earlier seems too good to be fortuitous. In some
parts of the world, such as in Namibia, these upper carbonate
deposits, which are always made of limestone (calcium carbonate) rather than dolomite (calcium magnesium carbonate),
reach hundreds of meters in thickness. Because carbonate deposition derives directly from the products of chemical weathering, the massive limestone cliffs of Namibia provide witness, not
only to the collapse of physical stratification, but to a massively
accelerated weathering regime during the hot aftermath of
Snowball Earth.[10]

It was Brian Harland's view that the Infra-Cambrian glaciation was something special, deserving of the epithet "great," and he has been spectacularly vindicated in the Snowball Earth hypothesis. Sedimentological and geophysical evidence for global ice cover is now overwhelming, which finally explains the enigma of glacially derived sediments lying on top of, and with barely any hiatus, apparently tropical oolite shoals and stromatolite reefs. The quandaries faced by so many geologists around the world like Tony Spencer, Bob Dalgarno, and Max Deynoux can finally be put to rest. The buildup of volcanic carbon dioxide explains how the earth finally escaped its icy fate, and the speed at which it did so. The return to a more normal global hydrological cycle, following ocean overturn, also explains why previously glaciated continents were now, once again, the sites of rapidly growing limestone platforms. Silicate weathering and the long-term carbon cycle slowly restored a degree of normality. Yes indeed, balmy shores and icy wastes, intimately juxtaposed.

But let's turn skeptic for a moment. I may have spun a convincing yarn, but a lot of what I have covered until now is inherently subjective, being one person's opinion on what amounts to only a tiny percentage of the remnants of an ancient seafloor. Much of what I have written invokes plausible, but perhaps not unique, explanations. Many of the claims I have made are also hard to quantify from the rock record alone, and, worse, some horrible circular reasoning exists, too. Do we really know that all of these deposits are truly of the same age? Many geologists, even now, dispute the synchronous nature of Cryogenian ice ages. Some geologists even dispute the uniqueness of the cap dolostone event itself. Perhaps cap dolostones relate not to a singular event, but to many similar events, taking place at different times in different places. Can we demonstrate beyond doubt that Harland's great glaciation began and ended simultaneously clear across the world? And are we talking about one great glaciation or many episodes of global cooling? In short, can we date glaciations with sufficient resolution, and can we go beyond dating to quantify the effects of the glaciation and its aftermath on the oceans and atmosphere? I think we can.

5

Clocks in Rocks

The resurgence of interest in global glaciation during the 1990s led to numerous tests of the super-greenhouse. High levels of carbon dioxide during two distinct Cryogenian ice ages (717–660 and 650–635 million years ago) have been confirmed using oxygen isotopes, while other geochemical evidence traces a surge in chemical weathering after the ice melted. The Snowball Earth hypothesis has been fine-tuned, but its basic principles survive unscathed.

I have been hiding a little secret until now and it is time to come clean. Although Brian Harland referred to the Great Infra-Cambrian Glaciation in the singular, it would be more correct to use the plural, for we now know that there were at least two specific ice ages during the Cryogenian Period. We can be certain about this because in some areas of the world, Canada, Namibia, and China, to name but a few, there are two distinct levels of glacial diamictite, separated by seemingly normal marine sediments. Douglas Mawson recognized two separate pulses of glaciation in Australia already in the 1950s, and it is the Australian names that have stuck. Today, most geologists refer to Cryogenian ice ages as being either "Sturtian" or "Marinoan," which refer to lower and upper rock series, respectively, that sit directly beneath Cambrian strata in South Australia.

The Nuccaleena Formation, you may remember, is the cap dolostone that marks the beginning of the Ediacaran Period and the end of the second, or Marinoan, phase of Cryogenian glaciation. The beginning of the Cryogenian Period is harder to spot in Australia because Sturtian glaciers scoured deeply into the older, underlying rocks, obscuring the transition into glaciation. We did take a look at the onset of the Sturtian ice age, however, in the previous chapter when we took a trip to Scotland. Although the end of the Sturtian glaciation in Australia is also marked by a thin dolostone cap, it is by no means of global significance. Geophysicists tell us that both ice ages occurred at low latitudes, and in Australia, Marinoan diamictites were deposited in apparent transitional contact with the underlying carbonate platform, called the Trezona Formation, confirming that both ice ages had abrupt beginnings, at least in the low latitudes.

When I arrived in Zurich in 1991 with my Scottish fieldwork still firmly in my mind, the number and order of late Precambrian glaciations was in a bit of a mess. Many people used Harland's approach, which assumed that diamictites were like time markers and could be correlated from one region to another. According to this scenario, all Neoproterozoic diamictite formations had exact time equivalents all over the world. However, there were major problems with this attractively simplistic approach. In many parts of the world, only one glacial deposit could be found. In others, there seemed to be three or more distinct deposits. Existing age constraints were often too

poor to determine for sure whether many of these glacial deposits were even of Cryogenian age. It turns out that scientists
were right to be skeptical, as we now know that ice sheets still
formed locally during the ensuing Ediacaran Period. As a consequence, many supposedly Cryogenian deposits have since
been reassigned an Ediacaran age and vice versa.

Many geologists over the years have proposed that cap
dolostones could help to solve this dilemma. These "pink cap
dolomites" were already widely used for correlation within
regions, but it was not universally accepted until recently that
they, unlike the underlying diamictites, did indeed represent
global time markers. Although there might not always be a cap
dolostone in a given region, the presence of never more than
one pink dolomite gave geologists a clue that Harland's approach had some merit, although skepticism and confusion
abounded. What was needed was an independent way to correlate ice ages on a global scale. The application of geochemistry
to correlation had been on the rise since the 1980s and eventually won the day. Advances in isotope geology, in particular,
during the last twenty years have been truly breathtaking, opening up an entire periodic table of opportunities. Isotopes are
now used for far more than merely establishing global tie-lines
and are increasingly employed to interrogate and quantify every
aspect of the super-greenhouse and plume-world hypotheses.

During the late 1980s and early 1990s, the Harvard geochemistry group, under the leadership of Stein Jacobsen, was
at the forefront of the isotope revolution. They had begun to
look at ways to quantify chemical weathering using a variety of
elemental systems and had begun to take particular interest in
the four naturally occurring isotopes of strontium, which have
masses 84, 86, 87, and 88. When I arrived in Zurich, another
hotbed of isotopism, I was fortunate to find a vacant strontium
isotope preparation laboratory and almost unlimited expertise

at my disposal courtesy of one of my doctoral supervisors, Rudolf "Rudi" Steiger, who had spent his career developing the field of isotope geochronology, which is the science of dating rocks. High-precision strontium isotopic analysis back then required a thermal ionization mass spectrometer, or TIMS for short, and ETH Zurich was, and still is, one of the world's top TIMS labs. Abandoning my field gear for the mask, clogs, and gloves of the "clean" lab, I learned how to tease various elements from their parent rocks, and how to use a battery of tests to evaluate whether purified elemental extracts reflected faithfully the chemical and isotopic composition of past oceans. Although now standard procedure, this convoluted and often intricate process is far from straightforward, and is different for each and every element and rock type. It can also be terribly disheartening. Only after many weeks of lab work, which itself followed a month or two of arduous fieldwork, is it possible to discover if hard-won samples are good enough to be of any use. I was one of the luckier ones.

The strontium isotope composition of seawater is precisely the same all over the globe but is known to alter its value as a result of changes in weathering, which is the dominant, although not the only, source of strontium to the world's oceans. Because some radioactive rubidium with mass 87 is always decaying slowly to strontium with mass 87, releasing a beta ray in the process, rocks undergoing weathering have significantly different amounts of ^{87}Sr relative to their "stable" sister isotopes, namely ^{88}Sr, ^{86}Sr, and ^{84}Sr, depending on the age and initial Rb/Sr ratio of the minerals in those rocks. Changes in the strontium isotopic composition of seawater, usually reported as the ^{87}Sr/^{86}Sr ratio, respond therefore to changes in global weathering regimes, but only on million-year time scales because the amount of strontium being delivered by rivers each year is dwarfed by the huge ocean reservoir they deliver into.

Today, the ratio is precisely 0.709160, give or take 0.000005, and it has reached that value only after myriad deviations like a staggering drunk. My colleague John McArthur at University College London (UCL) has shown that the ^{87}Sr/^{86}Sr ratio of seawater changes excruciatingly slowly, and never faster than about 0.000050 per million years over the past 500 million years, despite the many tectonic collisions, volcanic eruptions, and meteorite impacts that have taken place throughout that time. Nevertheless, because of the isotopic homogeneity of the world's oceans at any one time, strontium isotope stratigraphy can be used to date rocks that were deposited during times of major change. One of the steepest rises occurred between forty and eighteen million years ago and coincided with the rise of the Himalayan mountain chain. The Harvard group was the first to show that the rate of change in late Precambrian seawater was also impressively quick, suggesting major changes to the earth's weathering regime.[1]

Only a year into my PhD work, I had already been lucky enough to see firsthand the glacial deposits of both Scotland and China, the latter being supposedly the focal point of my study. My primary supervisor, Ken Hsu, had other ideas and suggested that I tag along on an imminent trip to the foothills of the Altai Mountains in western Mongolia (Figure 10). Martin Brasier of Oxford University organized the trip and was chair of the international working group tasked with establishing a global definition for the Precambrian-Cambrian boundary. Unlike many other paleontologists of the time, Martin had gained a convert's zeal in the interpretative power of isotopes. It was no easy task for our Mongolian hosts, led by the charming Dorj-iin Dorjnamjaa, to set up this wondrous trip of a lifetime. At the time, Mongolia was still suffering the aftermath of the fall of the Soviet Union. There were almost no fruits or vegetables in the largely empty markets, and we needed to take

Figure 10. Mongolian fieldwork. Paleontological study and debate
outside the communal *ger* (tent) between fossil sponge experts:
Max and Françoise Debrenne, Anna Gandin, Rachel Wood, and
Pierre Kruse on our joint field trip to western Mongolia in 1993.
(Photo courtesy of Rachel Wood)

all of our fuel with us for the 3,000-mile round trip. One of my
abiding memories of the capital Ulaanbaatar was of four of us
ordering a side salad and receiving a leaf of lettuce and a single
tomato cut into four portions. But the warm hospitality we experienced there still lingers in my memory, as do the horseback riding, wrestling, and throat-singing, tempered slightly by
less favorable memories of copious amounts of alcoholic mares'
milk in both fermented (*airag*, or *kumis* in Russian) and dis-
tilled (*arkhi*) forms.

The oldest sedimentary limestones we encountered in
Mongolia were of pre-Cryogenian age. They sat intriguingly
on top of thin slivers of green magmatic rocks that had once
been ocean crust, revealing how Mongolia had been stitched
together by the closure of many separate ocean basins. Indeed,

the whole of central Asia is a complex geological mess composed of bits of both oceanic and continental crust thrust together. The best way to envisage this "tectonic mélange" is to imagine how the western Pacific would look once the ocean there had disappeared completely through subduction. Some of these old rocks were the remnants of volcanic arcs that once stood proud, marking the slow passage of subducting lithosphere beneath. Long after the volcanism had ceased and the volcanoes had sunk beneath the waves, Cryogenian glaciation took hold, or so it would seem: the Mongolian diamictites fall into that awkward territory where glaciation is strongly suggestive but not conclusive. The team favored the glacial interpretation, but that begged the question of correlation. Was this the Marinoan, the Sturtian, or some other ice age yet to be discovered?

At the time, we hedged our bets and plumped for a Sturtian age, which was subsequently confirmed, courtesy of the hindsight gained from numerous studies around the world. The rocks overlying the Mongolian diamictites are not pinkish cap dolostones and show none of their characteristic features, but instead are dark, very finely laminated limestones that resemble other post-Sturtian deposits in Australia, Canada, and Namibia. Bringing the rocks back to Switzerland proved challenging, as I did not trust sending them by post. It took almost a week to get my now ninety kilograms of luggage back to Zurich via Ulaanbaatar, Moscow, and London. On my return, I lost no time setting about preparing them for analysis. The first step in the process is to cut the rocks to remove their outer surfaces and make thin slices for petrographic analysis. After a suitable part of the sample is chosen, a powder is made and dissolved in a dilute, weak acid, with the centrifuged solution passed through an ion exchange column, which I had calibrated to capture only strontium. The strontium-enriched portion is so tiny as to be invisible, so from this point everything proceeds on the basis of faith. Once evaporated and placed on a thin filament made of tungsten or rhenium, the sample is then heated by an electrical current under vacuum conditions. After temperatures well in excess of 1,000 degrees Celsius are reached, strontium ions are given off, deflected by an electromagnetic field, and collected in cups according to their mass/ charge ratio. The measured $^{87}Sr/^{86}Sr$ ratio results from numerous subsequent mathematical corrections and statistical analyses.

We quickly discovered that the Mongolian limestones were astonishingly rich in strontium, making them perfect candidates for isotopic work. The $^{87}Sr/^{86}Sr$ composition of the post-glacial Mongolian limestones changed from about 0.7067 to 0.7073 within just fifty meters of the pink-stained tops of the diamictites. At other times in earth history, such a massive change would have taken at least ten million years, or about as long as the entire non-glacial interval, which strongly suggested that post-glacial weathering rates were exceptionally high. Comparing our data with those pouring in from other regions of the world proved interesting reading over the following few years. Precisely the same values would be reported from other post-glacial sequences, while studies of much older pre-glacial limestones in Scotland, Greenland, and Svalbard showed similar $^{87}Sr/^{86}Sr$ values. The global isotopic fingerprint of these post- and pre-glacial limestones meant finally that these diamictites, all of which lacked an archetypal cap dolostone, could be assigned to the lower or Sturtian ice age and confirmed previous assumptions that the end of glaciation had been synchronous also for this earlier phase.[2]

Following the surge of interest in the Snowball Earth (or super-greenhouse) hypothesis after 1998, attention quickly turned to the younger of the two Cryogenian ice ages and its accompanying cap dolostones. But there was a major problem.

Marinoan cap dolostones everywhere were quickly found to have $^{87}Sr/^{86}Sr$ values all over the place, but generally much higher than is really possible for the global ocean. In Canada and Namibia, the crystal fan limestones overlying the cap dolostones told a more consistent story, revealing a pronounced upward trend, just like after the Sturtian ice age but starting much higher and rising more steeply, from about 0.7072 to 0.7080. The absolute values, once again, suggested little or no change across the glaciation. The rapid rise most likely reflects very high rates of chemical weathering and so was perfectly consistent with postulated super-greenhouse conditions after Snowball Earth. Interestingly, barite in Mauritania has an intermediate value, 0.7077, implying that the thin layer of condensed purple limestone draped over the platform interior of Mauritania was deposited at about the same time, but possibly more slowly, than the thick layers of aragonite fans on the Namibian shelf edge. Such consistency in $^{87}Sr/^{86}Sr$ values above the cap dolostones confirmed that post-glacial oceans were isotopically well mixed and increasingly dominated by terrestrial weathering input (Figure 11).[3]

The widely ranging and variable strontium isotope signature of the world's cap dolostones initially perplexed many observers but is perfectly consistent with rapid deposition and the predicted inhomogeneity of the meltwater plume. In the 1998 version of the Snowball Earth hypothesis, cap dolostones were considered to be the result of a huge input of alkalinity from rivers due to accelerated silicate weathering, but this interpretation is probably only part of the truth given the short duration of cap dolostone formation. Carbonate is more quickly weathered, especially as cold conditions enhance carbonate solubility, and there is the question of preexisting carbonate alkalinity in a super-greenhouse world. Whatever the case, in an expanding plume of meltwater, chemical and isotopic differ-

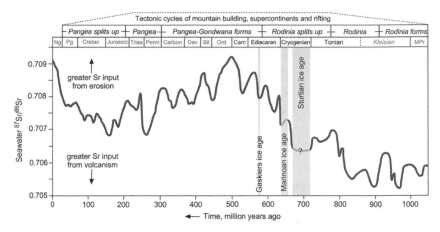

Figure 11. Seawater strontium isotopes. The ratio of seawater strontium-87 to strontium-86 fluctuates in response to grand tectonic cycles of erosion and magmatism. Seawater $^{87}Sr/^{86}Sr$ increases when mountains rise and decreases when supercontinents erode flat, or break apart to form new oceans. Sharp upturns after Cryogenian glaciations suggest that abrupt global warming and high carbon dioxide levels led to increased chemical weathering.

ences would be the norm for as long as physical stratification could persist. More recently, sophisticated sequential leaching experiments have been carried out on some cap dolostones with fascinating results. The new data show that strontium isotopic variation is not random, but that only some parts of the cap show anomalous values. In some cases, only the middle portions of the cap dolostone show the very high values typical of river runoff. Taken at face value, these new data imply enhanced mixing between the plume and the syn-glacial ocean water at the start and finish of cap dolostone deposition when, presumably, eustatic sea level rise outpaced the isostatic rise of the land.[4]

Despite such complications, the end result of ever more precise strontium isotope analysis has been to separate deglaciations into two distinct episodes. The earlier deglaciation was marked by the regional deposition of dark, laminated limestones with $^{87}Sr/^{86}Sr$ values rising from 0.7067 to 0.7072. Most glacial rocks from this earlier "Sturtian" glacial phase are marine, while evidence for a huge sea level rise due to ice sheet melting is lacking. Some think that this absence may be due to the extreme longevity of the Sturtian glaciation, during which continental ice sheets had time to sublimate, or creep, under force of gravity into the world's ocean basins. In high latitude areas, there appear to be no "cap carbonate" deposits at all, while deposition there seems rather ordinary. The later deglaciation was, by contrast, marked by global deposition of pale, pinkish cap dolostones with $^{87}Sr/^{86}Sr$ values rising from 0.7072 to 0.7080. Many glacial rocks of this later "Marinoan" phase are terrestrial and exhibit a huge sea level rise. Although one could speculate before about two distinct periods of global glaciation even without geochemistry, strontium isotopes provided an important independent line of evidence. However, the use of isotopes stretches far beyond mere stratigraphic correlation and allows sophisticated interpretations and even precise ages to be assigned to these two ice ages.[5]

Strontium is a relatively heavy element, so the mass differences between its four naturally occurring isotopes (with masses 88, 87, 86, and 84) are so small that its lighter isotopes react indistinguishably from its heavier isotopes. However, isotopic mass differences are much greater in lighter elements, causing redistribution of isotopes during an array of natural processes. This is because heavier isotopes tend to react more sluggishly but form stronger chemical bonds than lighter isotopes. Such "isotopic fractionation" is both an added complication and an aid to interpretation because the extent of fractionation can provide important clues to past conditions and processes on earth. Processes that fractionate isotopes can be biological, as when carbon is extracted from the air by plants and algae. Because photosynthesis favors the lighter carbon isotope, ^{12}C, which is less strongly bound to oxygen in the carbon dioxide molecule, organic matter is depleted in the heavier isotope, ^{13}C. Fractionation can also occur during physical processes, such as when, during evaporation, the lighter isotopes of oxygen (^{16}O) and hydrogen (^{1}H) tend to evaporate, leaving the heavier isotopes (^{18}O and ^{2}H) behind in lakes and seas. Both of these are examples of kinetic fractionation because they relate to the rates of these processes. In other words, more rapid evaporation would inevitably result in less isotopic discrimination. However, fractionation can also be related simply to the favoring of specific isotopes at equilibrium, whereby reactants and products rearrange isotopes in a predictable fashion determined by whichever distribution results in the lowest overall free energy. One of the first tests of the super-greenhouse hypothesis used one of the lightest of all elements: boron.

Boron is the element of choice for studies of ocean acidity. This is because the two isotopes of boron (^{10}B and ^{11}B) favor different borate species. At low pH, boron tends to exist as the isotopically heavy boric acid, while at high pH the isotopically light borate is more favored. When laboratories measure the isotopic composition of borate in a carbonate rock succession, any differences found between samples could therefore reflect changes in ocean pH. However, a geochemist's life is rarely that simple. Interpretations are in reality far less straightforward because the isotopic composition of boron in seawater can also change through time. Having two unknowns, seawater composition and pH, makes it impossible in practice to read pH directly from a boron isotope ratio in ancient samples. And that would be that, if it were not for the excellent work of Simone Kasemann, now at the University of Bremen in Germany. Simone, an isotope geochemist, learned about the Snowball

Earth debate through colleagues in Scotland, and realized that the extreme and sudden nature of Cryogenian deglaciation meant that there would be little time for the boron isotopic composition of the whole ocean to change significantly. In such cases, relative pH changes might still be detectable.

The first target was Paul Hoffman's stamping ground in Namibia, and the results were startlingly clear, although puzzling. Although it had been Simone's hope that the high carbon dioxide levels would imprint on the isotope record, neither the huge magnitude nor the transient nature of the effect appeared to make much sense. In Namibia, the cap dolostone and the transition to the overlying limestone witnessed an increase in acidity of almost two pH units, while the shift was very temporary indeed, beginning near the base of the cap and ending within just a few meters of its top. Kasemann continued her work at other localities in Namibia and even sampled sections in faraway Kazakhstan and China, but always with the same results. Sturtian deglaciation showed no such shifts, but Marinoan deglaciation always obliged, with ocean pH rapidly decreasing before recovering, either before the end of cap dolostone deposition or at a slight delay in the shelf break sections. High carbon dioxide levels explain the low pH shift, but only the limited buffering capacity of the brackish meltwater plume can explain such rapid acidification followed by amelioration of ocean pH. As rates of continental weathering went into overdrive, bringing masses of calcium and bicarbonate into the world's oceans, whole ocean pH would have rapidly recovered to near normal.[6]

The persistence of enhanced weathering long after deglaciation, as evidenced by strontium isotopes and likely driven by reinvigoration of the hydrological cycle, was a key prediction of the Snowball Earth hypothesis. Massively fluctuating pH in its immediate aftermath, as now evidenced by boron isotopes, was a key prediction of the plume world. But what about the super-greenhouse itself? Could we ever hope to find out what atmospheric conditions were like during the Snowball? Was there any evidence at all of high levels of carbon dioxide, for example? Finding any such smoking gun seemed an impossible hope, as how on earth could atmospheric conditions ever be captured in marine rocks or diamictites? Huiming Bao, then at Louisiana State University, believed he had the means to directly test the key prediction of Snowball Earth, high carbon dioxide levels, during glaciation. Huiming is a bubbly geochemist with surprising shocks of gravity-defying hair. Now back in China, he is engaged in setting up a suite of next-generation isotope laboratories in Nanjing.

Since the early days of stable isotope research in the 1950s, geochemists have focused largely on only two stable isotopes of oxygen, ^{16}O and ^{18}O, and ignored the third one, ^{17}O. There was a very good reason for this: the mass 17 variety is vanishingly rare. Oxygen with eight neutrons (^{16}O) is almost 500 times more abundant than oxygen with ten neutrons (^{18}O), which is in turn five times more abundant than oxygen with nine neutrons (^{17}O). Because isotopic fractionation predominantly depends on mass, any fractionation effect on ^{18}O (relative to ^{16}O) will be twice as great as on ^{17}O. Why bother making life difficult by trying to measure the abundance of such a rare isotope? This was the thinking, at least until technological advances made analyses easier, leading to the recognition that not all isotopic fractionation is exactly proportional to the mass differences between isotopes. In other words, some processes lead to a barely perceptible difference in the ratio $^{18}O/^{16}O$ than merely twice the change in $^{17}O/^{16}O$. Such tiny deviations from the expected pattern are termed mass-independent fractionation, or MIF for short, and measuring them with a mass spectrometer is quite a technical feat.

Mass-independent fractionation of oxygen isotopes was first discovered in meteorites and was assigned to photochemical dissociation of carbon monoxide during the early years of our solar system. Given that other photo-dissociation reactions, for example involving ozone, also lead to MIF in our own atmosphere today, there is a general presumption that MIF originates in chemical reactions between gases. Years of analyses have shown that terrestrial rocks, by contrast, show little in the way of measurable MIF, with one glaring exception. In test after test, Huiming Bao has shown that deposits formed during Cryogenian glaciation and barite formed in cap dolostones following deglaciation are characterized by extreme depletion in ^{17}O. In a series of ingenious papers, Bao has argued that the oxygen in the sulfate (in the barite) derived from the terrestrial weathering environment, and thus retained the signature of the atmosphere toward the end of the Marinoan ice age. According to Bao, MIF could only have come from interactions between oxygen, ozone, and carbon dioxide during the Snowball, while the extent of fractionation implies massively raised levels of that greenhouse gas.[7]

Bao's results have not been repeated for the Sturtian or, indeed, any other glacial episode in earth history, underlining the unique status of the catastrophic Marinoan aftermath. Consistent with the Snowball Earth hypothesis, high carbon dioxide levels, high global temperatures, and high chemical weathering rates presume glaciation to have been synchronous at its onset as well as at its close. But is this synchroneity really proven beyond all reasonable doubt? Strontium isotopes help with this question, as already argued, and so can carbon isotopes, as we shall see later, but fluctuations in seawater chemistry are rarely unique to one moment in earth history. Indeed, isotope stratigraphy can usually give only a relative age at best. Cynics don't give up easily, so to nail this one on the head what we need is a way to date rocks in an absolute sense in years. Fortunately, no field of geology has advanced more in recent times than geochronology, the science of directly measuring the age of rocks. Each year has seen increasingly precise age constraints that, at long last, leave no doubt whatsoever over the timing, duration, and synchroneity of Cryogenian glaciations.

When the Nuccaleena Formation was chosen to mark the start of the Ediacaran Period in 2003, this showed a considerable leap of faith, as its age was all but unknown. Moreover, no cap dolostones had been precisely dated anywhere in the world, which led to considerable criticism of this choice. There were no strontium isotope results from Australia for this interval either, and in retrospect it is surprising that the global body that decides such things (the International Commission on Stratigraphy, whose decisions need to be ratified by the International Union of Geological Sciences) was so brave. This was the first geological period to be established since the Ordovician Period in 1879. It was also the first boundary not to be based on fossil assemblages, and thus on biological evolution, but instead on a climatic and oceanographic event that in many scientists' eyes was neither synchronous nor particularly special. As with many naysayers, they had a point, and the best answer to such valid criticism is to get more data. Fortunately, science did not have long to wait.

At the same time that the golden spike was being hammered in, a team from the Massachusetts Institute of Technology (MIT) was beginning fruitful collaborations with expert geological teams from around the world to date Snowball Earth. On April 1, 2005, in the journal *Science,* Dan Condon and Sam Bowring of MIT, together with a team from the Nanjing Institute of Geology and Palaeontology (NIGPAS) led by Maoyan Zhu, published a precise age for the end of the Marinoan glaciation at 635.2 ± 0.6 million years ago (Figure 12). This

Figure 12. Ash bed sampling. Maoyan Zhu (Nanjing Institute
of Geology and Palaeontology, China) pointing out a greenish
volcanic ash bed sandwiched within glacial diamictite near the
base of the "Marinoan" Nantuo Formation (near Longbizui,
western Hunan Province). The ages of such volcanic events help
to constrain the age and duration of Cryogenian glacial episodes.
(Photo by Daniel Condon, 2002)

new age was—within error—indistinguishable from a similarly
precise age published in the September 2004 issue of *Geology* by the same MIT duo, together with Karl-Heinz "Charlie"
Hoffmann of the Namibian Geological Survey. The similarity
was remarkable because the former age derived from the post-
glacial cap dolostone in southern China, while the latter age was
for a level beneath the cap dolostone and within a unit bearing
glacial dropstones in northern Namibia. This sealed the case
for synchronous and rapid deglaciation following the second
of the Cryogenian Snowball events, and probably led to a huge
collective sigh of relief for everyone who had voted to ratify
the new period just a short time before. Later studies have only

further strengthened the case for globally synchronous deglaciation. But what made this incredible precision, better than
± 0.1%, possible for events that happened so long ago?[8]

It is generally very difficult to assign an exact age to ancient sedimentary rocks with any useful degree of precision,
by which I mean with a one-million-year error bar. The best
we can usually do is obtain relative ages based on their fossil or
isotopic contents. Radioactive decay can be used to date materials as long as we have accurate estimates of the half-lives
of the radioactive isotopes within them. However, this usually
does not work for sediments, as by their very nature they contain a lot of detritus that was inherited from the weathering of
much older rocks before being transported to the oceans by
rivers. Only maximum depositional ages can be obtained from
such detrital minerals, while if the sedimentary layering is cut
by magmatic intrusions, minimum depositional ages might also
be established. Very often, even exhaustive geochronological
studies provide too large a range between such maximum and
minimum ages to tease apart the time order of distant events,
while error bars on those ages may also be unacceptably high.
This is because in most radioactive series, it is very difficult
to distinguish how much of a given radiogenic isotope derives
from radioactivity and how much was already there. We encounter this problem when trying to date limestones directly
using the uranium-lead isotope decay series, which commonly
results in precisions no better than ± 2.5%, as I discovered
when I dabbled a little with lead isotope dating in Strasbourg
in the late 1990s, arriving at a perfectly useless age for the
post-Sturtian Mongolian limestones of 670 million years ago,
± 50 million years. Occasionally, however, we strike lucky in a
different way.

On very rare occasions, thin layers of volcanic ash may
be found sandwiched between sedimentary beds, and if we are

fortunate, we may find tiny datable crystals that formed during or shortly prior to the eruption. Zircon crystals are the best minerals for geochronology, and their ages form the backbone of the international geological time scale. This is because zircons incorporate radioactive uranium and thorium into their crystal lattices when they form, while excluding lead. This fact allows radiogenic lead to become the dominant form of lead in ancient zircon crystals. Three of lead's four isotopes (Pb-208, Pb-207, and Pb-206) are the radiogenic daughters of radioactive thorium-232, uranium-235, and uranium-238, respectively. There is an additional stable isotope of lead (Pb-204) that can be used to estimate the tiny amounts of "common" lead that were there to begin with. The existence of three independent decay series, which can be calibrated against each other, means that extremely precise ages can be obtained. Nowadays, events hundreds of millions of years ago can be dated to within a few tens of thousands of years. Such error bars of even better than ± 0.01% represent a marked improvement over what was possible even just a decade ago.

Within a few years of those ages being determined, Marinoan deglaciation had been dated in other regions of the world, too, including Australia and Mauritania, with the same results. The Sturtian ice age began at low latitudes around 717 million years ago, and it ended everywhere in the world synchronously at around 660 million years ago. Glaciation began again at low latitudes at about 650 million years ago, although this age is less well established, and ended everywhere at 635 million years ago. The speculation that Cryogenian glacial deposits formed at the same time and so can be used as a time marker has been spectacularly confirmed, although that simple fact does not help us understand it any better. The Cryogenian ice ages were clearly unlike anything before or since. It is hard, for instance, to compare them with the recent ice age, which stopped and started at various times during the Cenozoic Era. Following various ice sheet advances in the southern hemisphere, the northern hemisphere cycled in and out of glaciation after about six million years ago, before entering the current 100,000-year cycles of glacial and interglacial episodes over the past 2.5 million years that are driven by tiny but predictable differences in the earth's orbit around the sun. The tiny climatic effect exerted by the eccentricity of the earth's orbit gets magnified by positive feedbacks, which allow global temperatures to meander around a baseline climate. In comparison to this and other similar Phanerozoic ice ages, the Cryogenian Period represents a wholly different earth system, typified by runaway cooling, extraordinarily stable glaciation, and runaway warming unlike at any other time in earth history (with one possible exception over two billion years ago that we will get to later in this book). Extraordinary events require extraordinary explanations. But let's not get ahead of ourselves.[9]

One major outcome of this flurry of dating activity was entirely unexpected, at least in some quarters. For a long time, a few glaciologists, most notably Nikolay Chumakov, had suspected that some supposed Cryogenian glacial deposits were in fact considerably younger. Chumakov, a renowned geologist based at the Russian Academy of Sciences in Moscow, has been an unwavering promoter of Ediacaran glaciation for over half a century, publishing regional surveys and global syntheses in both Russian and English to near silence from the international geological community. It was the MIT team again that put Ediacaran glaciers firmly back on the agenda by dating the Gaskiers glaciogenic diamictite in Newfoundland at about 580 million years ago, fully 55 million years after the end of the Marinoan ice age, and smack bang in the middle of the Ediacaran Period. This age proved crucial for determining when animals evolved, as the deposits directly underlie the very first

large multicellular animal fossils. We will take a long, hard look
at these enigmatic fossils in a later chapter. Another glaciation,
the Luoquan glaciation of North China, appears, at about 560
million years ago, to be even younger, and there may be other
intervals of cold throughout the Ediacaran Period, for example around 610 million years ago. As we shall see in the next
chapter, this may not be surprising; we already knew from carbon isotopes that the Ediacaran and early Cambrian Periods
were times of extraordinary perturbations in the global carbon
cycle. Although clearly important climatic events, these glaciations were nothing like the Snowball ice ages, being merely
short-lived episodes of only regional glaciation at high latitudes and high altitudes. Nevertheless, they are not without
importance and their tremendous significance for the evolution of life and the biosphere is becoming clearer with every
passing year.[10]

High-precision geochronology has come a long way over
the past few years and will inevitably lead to further advances
as improved resolution permits us to test hypotheses of causa-
tion and correlation. Science progresses incrementally through
observation, prediction, testing, and refinement, ideally in that
order. But there is no such thing as a born scientist who rigor-
ously pursues the scientific method from cradle to grave. We
are all unfortunate prisoners of our limited personal experience, and those on the losing side of such arguments as Snowball Earth often continue to throw punches long after their
arguments have collapsed under the weight of evidence or
been honed by Occam's razor. Perhaps that's why scientific ad-
vancements are mostly community affairs, driven by curiosity
and ego in equal measure. Joe Kirschvink's shocking image of a
frozen planet trapped beneath a greenhouse blanket stemmed
directly from his own geophysical experiments, together with
the perceived wisdom that glaciers at the equator just had to

mean ice sheets everywhere else, too. Joe's idea and its reframing by Paul Hoffman and others made meaningful predictions that research groups all over the world were eager to test. Within a few years, groups of geologists, biologists, glaciologists, chemists, physicists, mathematicians, and climate scientists were tackling the hypothesis from all available angles. In the sense that it is testable, Snowball Earth is a good hypothesis. And good hypotheses evolve through strenuous testing. The supergreenhouse has so far stood the test of time. Synchronous deglaciation is no longer an unsubstantiated assumption, while evidence for global glaciation has never been stronger. However, the cap dolostone has changed its significance greatly. From its original role as the physical evidence for hugely increased silicate weathering rates, it has evolved to become the smoking gun for runaway deglaciation under a greenhouse sky.

Once we can accept something so crazily unhinged as global glaciation, we can perhaps finally conceive of the past as a truly different place. Abandoning the sacred creed of uniformitarianism can provoke exciting new questions. What could possibly have caused global temperatures to plunge like this? Could Snowball Earth ever happen again? What does it tell us about climate stability? Is it only a mirage of lucky happenstance that our planet has been habitable for so long? To answer such questions and march on toward the crux of my arguments about our own origins, we need to take a step back from this wacky interval in earth history and look at why our planet is habitable in the first place.

6
Broken Thermostat

The long-term carbon cycle acts as a thermostat, preventing climatic extremes through well-understood negative feedbacks. The Cryogenian Snowball Earth episode suggests that the earth's natural thermostat failed. Positive feedbacks may have helped push the world toward runaway cooling, but most suggestions are unconvincing. It seems likely that the global carbon cycle was fundamentally different back then.

The irrepressible Mr. Wilkins Micawber tottered between two poles of the human condition for many years before he ended up first in debtor's prison, and then pauper's exile in Australia. His summation:

Annual income twenty pounds, annual expenditure nineteen, nineteen and six, result happiness. Annual income twenty pounds, annual expenditure twenty pounds ought and six, result misery.[1]

What Charles Dickens described in his novel *David Copper-field* is common sense. When we fall short of the rent one week, the overspend must be corrected the next, lest we spiral into runaway decline. We all make sure to match our spending to our income so that, over time, average income will inevitably and increasingly approach average outgoings. Such dynamic equilibrium, whereby input equals output, leads to a condition called "steady state" because at that point your bank balance will have attained a constant size. Any subsequent non-steady state (earning less/more, spending more/less, higher/lower taxes) will inevitably cause your savings to either dwindle or accumulate. However, whether any new steady state can be achieved will depend on negative feedbacks that kick in whenever we fall short or feel flush. Negative feedbacks in the economy will prevent most of us getting rich long before the world runs out of money. The natural law of mass balance works the same way.

All natural cycles must obey the law of mass balance. If a given amount of material enters a reservoir, whether it be food moving through our guts, the air through our lungs, nutrients through the soil, or chemical substances through the oceans and atmosphere, then the same amount must also exit the reservoir, and at the same rate when averaged over a long enough time interval. If this were not the case, then the amount of a given substance retained by the system would continuously change in one direction, either disappearing entirely or rising to infinity. In most systems, neither of these extreme scenarios are reached before negative feedbacks kick in to bring the system back to stability. Scientists expect systems to respond to, and correct for, perturbations on different time frames, depending on the size of the reservoir relative to the net flux in and out. This simple ratio is referred to as the residence time. For example, the residence time of sodium in the ocean is many tens of millions of years because there is simply so much sodium in seawater relative to the comparatively paltry amount brought into the oceans annually by rivers.

At the other extreme, the residence time of carbon dioxide in the atmosphere is just eleven years, which is why we can see clearly the northern hemisphere seasons in the rise and fall of atmospheric CO_2 that takes place each year from autumn to spring. Over intermediate time scales of hundreds to thousands of years, carbon cycle imbalances are compensated for in the oceans, which either take in the excess or make up the deficit in atmospheric CO_2. However, being reversible processes, neither respiration and photosynthesis nor carbonate precipitation and dissolution represent permanent sources and sinks of carbon dioxide. The residence time of carbon in the entire surface environment is on the order of a hundred thousand years. In other words, it could take as long as a hundred thousand years of volcanism and weathering to build up the amount of CO_2 stored in the earth's biomass, oceans, and atmosphere combined. Only over such long time scales might we expect negative feedbacks to balance the sources and sinks of carbon dioxide completely.

As we saw in chapter 4, the long-term carbon cycle—and so climate—is believed to be regulated by rock weathering. It can achieve this impressive feat because weathering, a major sink of CO_2, is temperature-dependent, and so any shifts in global climate ought to be compensated for by opposing changes to the strength of the greenhouse effect. Despite the operation of this classic negative feedback, it appears that temperatures fluctuated wildly during the Neoproterozoic Era, with seemingly abrupt shifts from one climate state to another. The earth's greenhouse blanket seems to have failed repeatedly and irrevocably during that time, allowing glaciers to encroach so close to the equator that the entire globe became enveloped in a shroud of ice for millions of years. During that time, greenhouse gas levels rose inexorably with little hindrance so that once glaciation ended, super-greenhouse conditions followed. Moreover, the resultant massively elevated weathering rates seem not to have resulted in immediate cooling either. Such fluctuation without moderation flies in the face of conventional scientific wisdom and suggests the existence of hitherto unsuspected positive feedbacks, which allowed climate to pass beyond crucial tipping points to reach an entirely new normal. Why extreme amplitudes of climate change occurred throughout this time in earth history forms the topic of this chapter and may help us to understand why the Neoproterozoic Earth System, rather like Micawber's purse, lacked resilience, or the ability to bounce back after perturbation.

A man from a bygone age who understood the "normal" economy of natural cycling was James Hutton. Despite being known as the "Father of Geology," Hutton was just as suited to the study of cycles as he was of rocks. The doctoral thesis he completed in the Netherlands centered on the circulation of blood around the body. In his seminal work on geology, published many years later, one can see the influence not only of his earlier work on blood but also of the Enlightenment zeitgeist in his writing, "I see a circulation in the matter of the globe, and a beautiful economy in the works of nature." Echoing these sentiments, his friend David Hume, the well-known philosopher, declared that "a continuous circulation of matter . . . produces no disorder. A continual waste in every part is incessantly repaired; the closest sympathy is perceived throughout the entire system." In other words, cyclicity begets a dynamic equilibrium, which arises spontaneously by a process of restorative "negative feedback." A good example of this was proposed by another great Enlightenment thinker: Adam Smith maintained that a perfect equilibrium between supply and demand would emerge of its own accord in a free market economy. Clearly, balance and counterbalance were all the rage around the turn of the eighteenth century.[2]

It could be that James Watt and his steam engine were responsible for turning people's attention to negative feedbacks. In his 1788 version of the centrifugal governor, the speed of rotation is kept constant by two heavy balls that spin around an upright frame. Whenever the frame spins too quickly, the centrifugal force raises the balls, and in so doing slows the motion by closing the oil or steam inlet valve. This simple device, borrowed from windmills, exercised many of the great minds of the subsequent nineteenth century as well. In his address to the Linnean Society in 1858, Alfred Russel Wallace wrote that the evolutionary principle, which Darwin published in full only the following year, works like the centrifugal governor "which checks and corrects any irregularities almost before they become evident." In engines, organisms, the economy, and our planetary environment, negative feedbacks spontaneously govern the stability of complex systems.[3]

The negative feedback believed to govern climate, and thereby planetary habitability, would not be completely expressed until a classic paper in 1981 by Walker, Hays, and Kasting, which postulates a mutual sensitivity of both silicate weathering and climate to carbon dioxide. It seems ironic now that when James Hutton announced his new "Theory of the Earth" to the world, he did so in a lecture titled "Concerning the System of the Earth, Its Duration and Stability" that was given by his close friend Joseph Black. It was Black who first divined the acidic nature of carbon dioxide, which he determined to be the major constituent of both shells and limestone ($CaCO_3$), while at the same time being produced from burning coal. Black called his new gas "fixed air," and soon the great chemist Joseph Priestley showed how fixed air could be released from limestone simply by dropping acid on it. His seminal paper

"Impregnating Water with Fixed Air" presented carbonic acid, the main agent of silicate weathering, for the first time.[4]

Black knew that given enough time, rainwater, which is, after all, just dilute carbonic acid, could weather entire mountain chains to the ground, and Hutton spent much of his life proving it. Hutton's marvelous insights reached beyond geology to encompass the cycling of all materials, most importantly nutrients to sustain life. In words strangely prescient of James Lovelock's Gaia theory, which views the earth in terms of a self-regulating, self-sustaining organism, Hutton wrote that the earth "is a compound system of things, forming one whole living world. . . . The matter of this active world is perpetually moved, in that salutary circulation by which provision is so wisely made for the growth and prosperity of plants, and for the life and comfort of its various animals." Snowball Earth would have come as a terrible shock to Hutton and his contemporaries, whose ideas of perpetual stability would have been sorely put to the test. He famously wrote that the earth's enormously long pre-history has "no vestige of a beginning, no prospect of an end." However, as Darwin and Ebelmen were among the first to realize, we live in a directional world in which no two cycles repeat each other precisely and in which even the nature and direction of feedbacks can change. Despite this, the geologist's dislike for catastrophism and preference for stabilizing, regulatory feedbacks, rather than destabilizing, positive ones, remains undiminished to this day. However, if we want to dig into the reasons for unprecedented events such as Snowball Earth, we might need to discard the uniformitarianism rulebook, or at least leave it to gather dust for a time in a drawer.[5]

Carl Sagan, the peerless popularizer of science, public broadcaster, novelist, and critical thinker, summed up his search for extraterrestrial life with the memorable words, "If we are alone in the universe, then it is an awful waste of space." In

Sagan's opinion, given both the vastness of the universe and the enormity of time, the improbable becomes the inevitable. This might also be a suitable mantra for geologists who deal with similarly huge spans of time and similarly improbable, but inevitable and catastrophic, events like meteorites, eruptions, earthquakes, and tsunamis. In a lifetime sadly cut short in 1996, Sagan continually encouraged governments and individuals alike to think outside the box. He had a fascination that I share for human irrationality and actively promoted research into its origins. His lecture course on "critical thinking" was given exclusively to just twenty handpicked students that were selected from their written essays. Such elitism is almost a shame, as we could all benefit from such a course, perhaps in inverse relation to our ability to write a decent essay on the subject. Sagan enters our tale in 1972, the year following his appointment as the chair of planetary science at Cornell University.[6]

As a physicist, Sagan was aware that stars give off more heat as they age. This heating up is an inevitable result of the nuclear fusion reactions that take place in the sun's core. A modern estimate for our own star is that it gives off at least 25% more heat now than it did when the world's most ancient microbial carpets were still slimy to the touch. Planet Earth supports life with the same genetic template today as it did back then and presumably for all the time in between, which means that its climate has never once ventured outside life's comfort zone, defined largely by the need for liquid water. This is commonly called, following Sagan's lead, the "Faint Young Sun Paradox." In one of their earliest joint contributions on the Gaia hypothesis, Lynn Margulis and James Lovelock took Sagan's idea on board when they wrote, "The probability that, by accident, the temperature has for 3.5 billion years, without exception, followed the straight and narrow path optimal for surface life seems unbelievable." Their conclusion that "life must

actively maintain these conditions" echoed James Hutton's notion of a homeostatic earth 200 years earlier.[7]

Whatever one thinks of the Gaia theory, and we will revisit it again toward the end of this book, it is intriguing (although for some perhaps only highly fortunate) that life has
maintained its hold and surface temperature has remained quite
muted despite the fact that the sun has relentlessly been turning up the oven knob over billions of years. Clearly, some other
climatic forcing mechanism must have changed—incrementally—to compensate for the ever strengthening sun. There
are only two factors that can achieve such a magnitude: the
albedo, or reflectiveness of the earth's surface, and the greenhouse effect. Carl Sagan thought that levels of ammonia were
higher in the distant past to compensate for the Faint Young
Sun, and some have preferred methane. However, for various
sound reasons, it is carbon dioxide that must get the most credit
for this balancing act, presumably via the canonical climate—
weathering negative feedback.

It was another of the Eiffel Tower grandees, Jean-Baptiste
Dumas, the man who weighed the elements one by one during
the first half of the nineteenth century, who in 1841 first postulated that carbon dioxide (and nitrogen) levels were higher
in the earth's distant past, in the first ever description of the
carbon cycle. Four years later and more than a century before
it was rediscovered, Jacques-Joseph Ebelmen wrote an almost
complete description of the long-term carbon cycle. To paraphrase in translation from the French, he wrote, "I see in volcanic phenomena the principal cause that restores atmospheric
carbon dioxide that has been removed by the decomposition
of rocks. Plant roots can accelerate the weathering of silicate
minerals with which they are in contact, releasing carbonate
ions that end up being deposited as the shells of sea creatures."
Mixing the ideas of Sagan, Dumas, and Ebelmen together, in a

smorgasbord across the ages, we can surmise that during Precambrian times, when the sun was fainter, higher carbon dioxide levels could have been sustained because of the absence of
higher plants that accelerate weathering. As plants became ever
more efficient weathering agents, the greenhouse effect weak-
ened in turn, leaving the earth's climate progressively more vulnerable to carbon cycle perturbations.[8]

Since the mid-1840s, only the discovery of plate tectonics,
fully 120 years later, represents a major advance in our understanding of the long-term global carbon cycle. We now know
that carbonate only sometimes makes its way back to the surface via the Huttonian process of purely vertical movements
and erosion. In recent decades we have been able to confirm
that the seafloor is dynamically changing, while uplift is primarily driven by horizontal, not vertical, forces. The oldest part
of the modern ocean floor formed less than 200 million years
ago, and less than 5% of earth history has passed since its
deposition. Most volcanic carbon dioxide today is therefore
the product of another great cycle, a tectonic cycle whereby
fossilized organic remains and carbonate grains and shells give
up their volatile carbon as the ocean lithosphere (crust and
uppermost mantle) creep beneath the continents at a rate of
centimeters per year. Without this "subduction" there could
be no long-term carbon cycle, no atmospheric regulation, and
presumably far more limited possibilities for life on earth.

Because the carbon dioxide that pours out of volcanoes
should roughly match the removal of carbon dioxide by silicate
weathering (and subsequent carbonate deposition), any slack
ought to be taken up by the other long-term carbon sources
and sinks, which are the weathering and burial of organic matter, respectively. On time scales of a hundred thousand years
or longer, all these sources and sinks ought to be perfectly in
sync, and yet Snowball Earth happened. What made the earth's

climate system less resilient during those times? What went wrong with the earth's planetary-scale thermostat? There are a number of interesting theories.

The first idea involves methane and invokes the spirit of an earlier series of glaciations more than 1,500 million years before. Although far less well preserved than the Cryogenian deposits, geophysical evidence suggests that at least one of these earlier "Paleoproterozoic" ice ages may also have been a global event, so they are commonly, but not universally, referred to as the first Snowball Earth. Despite their vast age, there is less dispute about their cause, which is universally linked to the Great Oxidation Event, or GOE, which represents the moment when free oxygen first appeared in the atmosphere on a global scale, where it could react with methane, collapsing the earth's greenhouse blanket. Could something similar have happened during the Neoproterozoic Era? Possibly, but, to be frank, there does not appear to be a lot of evidence in favor of high levels of methane in the pre-Cryogenian atmosphere. Before the GOE, by contrast, not only would high levels have been expected due to the lack of atmospheric oxygen, but isotopic evidence in the form of organic matter, highly depleted in the isotope carbon-13, was abundant. Depletion of ^{13}C is a reliable geochemical fingerprint for methanogenesis but is entirely absent before both of the Cryogenian ice ages, neither of which have specifically been associated with major atmospheric oxygenation. This is not to say that methane did not play any contributing role at all, and we shall return to that role a little later.

Another contributing factor might have been the evolution of more complex soil biota. As mentioned briefly above, land plants evolved in tandem with a progressively weakening greenhouse blanket. Could a major evolutionary event have triggered the glaciations? At some time during the Proterozoic, it seems likely that the continents would have experienced their first major "greening" before the evolution of higher land plants. Mosses, liverworts, and modern vascular plants with leaves and roots all came much later during the Paleozoic Era, but lichen, symbioses between algae and fungi, could possibly have emerged during the Tonian Period, which immediately preceded the Cryogenian Period. They would have sequestered carbon dioxide more effectively not only by accelerating silicate weathering but also by increasing organic burial. This is because lichens are very good at leaching phosphorus from rocks, thus fertilizing the oceans. Organic steranes, likely products of red and green algae, become much more common during and just before the Cryogenian ice ages, lending some weight to this idea, although both these groups first left fossil remains more than 300 million years earlier. Similar biological innovations are associated with other glaciations in earth history, in the late Ordovician, Carboniferous-Permian, and even our current ice age, which began around the time that C4 plants, like grasses, spread around the planet. Although a neat idea, it is a little difficult to know whether the evolution of such photosynthetic carbon concentration mechanisms (CCMs) were a cause or a consequence of lower carbon dioxide levels. Notwithstanding that objection, none of these other glaciations even approached global extent, so something else must have been at play here.[9]

An intriguing idea often referred to as the coral reef hypothesis has also been invoked to explain both current glacial-interglacial cycles as well as the Cryogenian deep freeze and relies on the simple fact that glaciation must lead to a global sea level fall. Such sea level changes are referred to by geologists as eustatic changes, or eustasy, as opposed to isostasy, which refers to the regional flexing of the earth's ductile interior due to the shifting weight of its overburden, causing only a relative change in sea level. When sea level falls due to the expansion of ice caps, recently formed limestone coral reefs are exposed to dissolution by the rain. This causes a transient decrease in atmospheric carbon dioxide, which in turn causes further cooling and ice cap expansion in a cycle of positive feedback that continues until carbonate precipitation and the silicate weathering thermostat eventually rein in the runaway cooling after thousands of years. This and other positive feedbacks likely contribute to the otherwise puzzling fact that orbital cyclicity is simply too weak to explain the magnitude of climate change between glaciations over the past 2.5 million years. During Precambrian times, however, the significance of the "coral reef" effect may have been far greater because, before the evolution of oceangoing calcareous plankton, almost all carbonate would have been deposited in shallow marine environments. Although this is a great idea, foraminifera and other microscopic shelly organisms that make up the "calcareous ooze" covering the deep seafloor today (and the chalk of the white cliffs of Dover) only became important much later during the Mesozoic Era. Why, then, did no ice age during the intervening 500 million years turn into a Snowball event?[10]

More recently, the search for an explanation has turned to the major sources of carbon dioxide today. Different groups of researchers have assembled convincing evidence that the Cryogenian Period marked a relative lull in volcanic outgassing. This is because it was a fairly inactive time for tectonic plate collisions and arc volcanism in particular. Statistical analyses of mineral populations, most notably zircons, have shown that long-term climate trends have tended to covary with global volcanic activity, so tectonic quiescence could certainly have been a prerequisite for some cold spells. Yet it is hard to place the onset of the Sturtian ice age with a lull in outgassing; rather, the reverse is more likely. For much of the previous 200 million years, the world's continents had been welded together, following successive tectonic collisions, to form a supercontinent, Rodinia. But that all changed in the run-up to the Snowball glaciations, which began after Rodinia began to rupture, releasing huge volumes of basaltic lava and gas, including carbon dioxide. Prolonged wearing down of supercontinents can also lead to lower erosion rates and therefore lower oxidative weathering of sedimentary organic carbon, which is the other major source of carbon dioxide today. However, the same tectonic scenario repeated itself following the rifting of the supercontinent Pangaea, 500 million years later, so it is hardly unique, while that later interval is strangely marked by global warming rather than cooling. So, yes, lower outgassing rates may have contributed to the earth's vulnerability to glaciation, but their importance and uniqueness are questionable at best.[11]

For the sake of completeness, we should also think about the major sink for carbon dioxide today, organic burial. Suggestions that larger, faster-sinking phytoplankton evolved during the Tonian Period have been around for decades, while new soil biota may have helped to produce clay minerals that acted as ballast for the sinking organic matter. I could go on, but you already must realize that there are many ideas out there for why Snowball Earth happened, and most of them are not unique to this particular time period. We have already listed atmospheric methane crisis, photosynthetic innovations, the coral reef hypothesis, low carbon dioxide outgassing, and high carbon dioxide removal by organic burial, but none of these phenomena explain why a system in equilibrium experienced such major change in the absence of other drivers. Do biological innovations, for example, really drive change as soon as they appear, without a permissive environment to allow their evolutionary success? And we still have not worked out why any of these phenomena would have broken the back of the silicate weathering thermostat. How could such relatively unexceptional phenomena make it go AWOL? We need to dig still deeper.[12]

One of the first explanations for Snowball Earth arose from the emerging consensus that Rodinia was primarily a low-latitude supercontinent. In chapter 4 we visited the geophysical evidence that glaciation took place even at equatorial latitudes. But as more evidence amassed to secure this extraordinary claim, it quickly became apparent that a surprisingly large portion of the world's continental crust must have been at low latitudes, far more than today under the present continental configuration. It also became clearer with time that some of these cratons were beginning to split apart due to early stages in the rifting of Rodinia, leading to what has been referred to as the "fire and ice" hypothesis. Taken together, these two facts meant that sea ice could potentially encroach upon lower latitudes without forming thick ice caps like today, up until the point of no return given the runaway albedo of expanding sea ice. Recently erupted basalts in those continental rifts would have continued soaking up carbon dioxide because their tropical location would have been the last place on earth to experience the global cooling. Therefore, the unusual continental configuration separated the two key parts of the thermostat, silicate weathering and its cooling effect. More recently, it has been suggested that the high phosphorus content of those basalts further exacerbated the cooling due to fertilization of the oceans and organic burial. This is still a really good idea, and probably the only one that could potentially break the back of the weathering thermostat, which, if the idea is correct, is looking less like a thermostat and more like a mirage, only visible when we want to see it.[13]

There is a problem, however, with this purely tectonic happenstance model, and that is the carbon isotope record. Not surprisingly for a period when climate was going haywire, the carbon cycle viewed through the lens of the carbon isotope record was going equally crazy. The carbon isotope composition of seawater during the Cryogenian (and subsequent Ediacaran) Period is simply off the scale when compared with most other times in earth history. Although widespread rifting and continental configuration are undoubtedly important for the global carbon cycle, they do not seem to be unique enough to explain—on their own—both the extraordinary climatic and isotopic events of that time. In order to paint a fuller picture, while keeping all of the above possibilities in our minds, we need to abandon our preconception that the present provides the key to the past. Why should we use the modern carbon cycle to understand the workings of the Precambrian Earth System? In other words, can we be sure that the Neoproterozoic carbon cycle even during non-glacial times was anything like our modern carbon cycle? Can we even be certain that the residence time of carbon in the biosphere was a mere 100,000 years? If it were much longer, then imbalances could be sustained for longer, too, and render the biosphere less resilient against sustained perturbations. The earth was different back then, and to see quite how different we need to delve more deeply into our planet's history well before the earth froze over.

Geologists divide earth history into four huge chunks of time called eons. The first part, the Hadean, left almost nothing for posterity in the way of a rock record, which really kicks off about four billion years ago with the onset of the Archean Eon. The Proterozoic Eon lasted for about two billion years starting at 2.5 billion years ago and ending with the Ediacaran Period that followed the Cryogenian glaciations. We live in the ensuing Phanerozoic Eon, which is characterized by a more modern-seeming world, populated by animals and plants. Appropriately enough, Proterozoic denotes "early life," while

Phanerozoic refers to "visible life." What sets the Proterozoic world apart from the previous Archean one is oxygen, with the GOE (Great Oxidation Event) roughly marking the boundary between these two eons. Although there was virtually no free oxygen in the Archean atmosphere, the Proterozoic atmosphere was continuously oxygenated and so, we believe, was the surface (but not deep) marine environment. The Proterozoic Eon therefore marks a transitional situation between the modern fully oxygenated world and the Archean fully anoxic world. This distinction could provide a very important clue to why the Proterozoic carbon cycle acted differently from the Phanerozoic carbon cycle.

The GOE left its mark in several features of the rock record. For example, it was after the GOE that the first red beds appeared, the color deriving from oxidized, ferric iron (Fe III). Volcanic lavas have significantly higher amounts of ferric iron from this point on in earth history, showing that the earth's crust began to rust on land but not yet on the seafloor. Sedimentary rocks are entirely devoid of detrital pyrite grains (iron sulfide) after the GOE, showing that oxygen levels never returned to their Archean minima again. The GOE had many consequences for life and its environment, which we will revisit in a later chapter, but for now it suffices to recall that the event represented the surpassing of a threshold, whereby oxygen sources, largely oxygenic photosynthesis coupled to organic burial, were able to overwhelm oxygen sinks. After the GOE, at least one major oxygen sink, namely the weathering of sulfide minerals like pyrite, became saturated, allowing rivers to carry sulfur as oxidized sulfate (SO_4^{2-}) ions into the Proterozoic oceans. Similarly, rusted iron (II) could now be transported into the oceans as oxidized iron (III), but only in solid form as Fe^{3+} ions are largely insoluble. As a result, iron particles carried by rivers must have changed irrevocably following the GOE from iron sulfides to iron oxides. Once in the anoxic oceans, however, a portion of the oxidized sulfate and insoluble iron would have been reduced back again to sulfide and soluble ferrous iron, respectively, allowing pyrite to re-form following a series of reactions known to be mediated and catalyzed by microbes. At steady state, oxygen consumed by pyrite weathering is returned by pyrite burial, and this plays a key role in balancing the global oxygen budget even today. The other major part of the oxygen budget relies on a similar balancing act between organic carbon oxidative weathering and burial.[14]

In today's earth system, oxygen gets consumed by a combination of reductant reservoirs. As well as pyrite, many other sulfide or ferrous iron minerals and volcanic gases are also readily oxidized in our surface environment, but the largest oxygen sink is organic weathering. About 20% of all carbon in rocks is ancient fossil carbon that was once part of a living organism and still persists as coal, oil, or disseminated organic matter of various types. Any organic matter, when exposed to the elements, will eventually, given enough time, undergo oxidation to carbon dioxide in our fully oxygenated rivers, oceans, and atmosphere. However, even today, not all fossil carbon will be oxidized. In areas of rapid erosion, such as the Himalayas, where rocks may be exposed to weathering for short periods only, relatively refractory or inert carbonaceous compounds may be transported rapidly away in particulate form to the sea, where they will begin a new rock cycle relatively intact. How much organic matter ends up as inert detritus is a function of exposure time and oxygen levels. In other words, oxidative weathering, which is a major source of carbon dioxide to the atmosphere and a major sink for oxygen, is an example of kinetically limited weathering. On the Proterozoic earth, with relatively low levels of atmospheric oxygen, kinetic limitation of oxidative weathering would have been a key buffer against both oxygenation and climate change. The carbon isotope record shows us this very neatly.[15]

It has long been recognized that the period between the GOE and the Cryogenian Period, and especially between 1.85 and 0.85 billion years ago, is characterized by very muted variability in the carbon isotope composition of seawater. Professor Martin Brasier of Oxford University jokingly referred to this interval as the "boring billion," apparently inspired by an article he was reading about the tendency of billionaires to be rather boring people. Carbon isotope values, or $\delta^{13}C$ values, represent the proportion of the isotope carbon-13 relative to carbon-12, with higher values reflecting greater amounts of the heavier carbon-13 isotope. Today, dissolved ocean carbon has a $\delta^{13}C$ value of about zero per mil, which means it has the same $^{13}C/^{12}C$ ratio as the international laboratory standard. The standard geochemists use is the fossilized shell of a belemnite, a type of squid-like animal that once swam in the warm Cretaceous seas around California. Carbon isotope values have remained about zero per mil over most of the earth's history, but there have also been minor deviations away from the norm to keep things interesting. Such excursions (or anomalies) are usually interpreted as a reflection of organic burial rates, whereby higher values imply higher rates of burial. This is because organic matter is enriched in ^{12}C, so when more of it is buried, the oceans become isotopically heavier. If organic matter is buried, then the oxygen released during photosynthesis can stay in the atmosphere, which means that higher $\delta^{13}C$ values are generally associated with higher rates of oxygen release (Figure 13). However, during the boring billion there are precious few such blips, hence the name. Values in the Tonian Period get slightly more interesting, with both positive and negative excursions that become progressively larger toward the Cryogenian Period. The Cryogenian, Ediacaran, and early Cambrian records

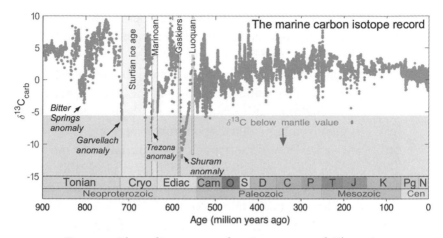

Figure 13. The sedimentary carbon isotope record. The ratio of carbon-13 to carbon-12 in marine carbonate rocks and fossils ($\delta^{13}C$) has varied over time in response to changing ocean chemistry. High and low values correspond to times when organic burial and carbonate burial dominated the ocean-atmosphere system, respectively. The Tonian-Cryogenian-Ediacaran interval (about 800 to 550 million years ago) is characterized by short episodes of extremely low values, suggesting huge perturbations to the global carbon cycle, with likely consequences for climate (see Sturtian, Marinoan, Gaskiers, and Luoquan glaciations) and the earth's oxygen budget.

are all rather wild and have proven to be really hard to decipher. Fortunately, we do not need to do that until chapter 10.[16]

The simplest and best explanation for the boring billion was given to us recently by numerical modelers based at the University of Exeter and the University of Leeds. They proposed that the kinetic limitation of oxidative weathering was so great during most of the Proterozoic Eon that very large amounts of organic matter remained pristine throughout the entire weathering process. This means that the usual way of explaining the carbon isotope record simply did not apply to the boring billion. To illustrate this point, let's imagine that atmospheric oxygen levels increased on a kinetically limited planet. This would have had the effect of oxidizing more fossil carbon, in so doing releasing more carbon dioxide into the earth system. Organic matter is enriched in the lighter isotope, ^{12}C, so oxidizing more of it would decrease the isotopic ratio ($\delta^{13}C$) of dissolved inorganic carbon in seawater and marine carbonate rocks. However, organic carbon burial is the most likely cause of the increase in oxygen in the first place, and, as we saw above, organic burial increases the isotopic ratio. The clever modeling of Stuart Daines, Ben Mills, and their mentor, the exceptional polymath Tim Lenton, shows that the resulting negative feedback would make it hard for the Proterozoic carbon isotope record to be anything but uneventful. The recognition that a lot of organic matter would have escaped oxidation carries the profound implication that atmospheric oxygen levels remained suppressed by this oxygen capacitor throughout much, if not all, of the boring billion. However, both carbon isotopes and climatic extremes tell us that the earth system became increasingly volatile throughout the Neoproterozoic Era. What could possibly have happened to push it so far out of kilter? Clearly, something else was going on here, and it has to do with the fate of the slowly reacting, kinetically limited organic pool.[17]

Another trio of mathematically minded geoscientists led by Dan Rothman at Harvard University had already proposed a solution in 2003. They suggested that the residual, unoxidized organic matter simply built up in the oceans to extraordinarily high levels. A quick back-of-the-envelope calculation suggests that the Proterozoic ocean would have been like one enormous peat bog, with clear waters presumably only at the surface exchanging oxygen freely with the fully oxygenated atmosphere.

In their model, the isotopic excursions that ended the boring billion represent the episodic depletion and eventual exhaustion of that organic carbon reservoir. An interesting corollary of this speculation is that the unreacted organic matter would also have acted as a buffer against climatic as well as oxygen changes because the drawdown of oxygen into the ocean during cooling—gases are more soluble in cold water—would oxidize more organic carbon, producing carbon dioxide that would warm the climate up again. Some climate modelers have even suggested that something like Snowball Earth would simply be impossible in the presence of such a powerful capacitor, which likely waxed and waned following the GOE. This negative feedback is potentially an explanation for why there is no evidence for glaciation of any kind during the boring billion of only muted isotopic variations, but implies a fundamentally different feedback from the silicate weathering thermostat studied the world over by geology students. Perhaps the earth's distant past really was a different country.[18]

Importantly, negative anomalies (low $\delta^{13}C$ values) precede both the Sturtian and the Marinoan glaciations, suggesting that this crucial climate buffer may have become severely compromised at those times. In both cases, it is the recovery from extremely low carbon isotope values that directly coincides with the first evidence of low-latitude glaciation in Scotland around 717 million years ago and in Australia around 650 million years ago. The demise of such a key climate buffer might indeed be enough to explain Snowball Earth, as it would have rendered the climate system more vulnerable to other climate forcing mechanisms such as basalt weathering, the coral reef effect, along with any weakness in the silicate weathering thermostat. However, in order to get a fuller picture of what was going on with the carbon cycle during these times, and how it might

relate to biological evolution, we need to step forward in time
to the aftermath of Cryogenian deglaciation and the ensuing
Ediacaran Period. Although the pre-Marinoan anomaly is very
large, it is not the biggest negative anomaly of all, which occurred toward the end of the Ediacaran Period. Although the
Ediacaran world witnessed episodic cooling and even glaciation, there were no ice ages of global extent. Intriguingly, however, the Ediacaran anomaly directly precedes regional glaciation and an amazing blooming of animal life on earth. The
anomaly itself is associated with the radiation of aerobic (oxy-
gen-requiring) organisms of great size and complexity, while
the recovery from the anomaly witnessed the emergence of
the first reefs, the first shells, the first trails, the first burrows,
and the beginnings of what is known more commonly as the
Cambrian explosion of animal life. Life was getting large, powerful, and mobile, and escalating very quickly courtesy of a
muscular arms race that would be familiar to us today; a world
in which big things chase and eat little things, which in turn
learn all manner of survival strategies.

It would be easy to imagine that complex life began
straight after Snowball Earth. Indeed, imagine how calamitous
a change it would be were today's world suddenly to plunge
into a permanent deep freeze. It is hard to envisage many modern life forms surviving such an existential crisis. And yet survive it must have, as we know that life has an unbroken genetic
chain stretching back to its Archean origins. Even more surprisingly, it would appear from comparing modern animal
DNA with the fossil record that we owe our very origins to
whichever ancestral organisms managed to adapt best to those
extreme conditions. We now think we know how the thermostat broke and that it was not even entirely the same thermostat that keeps conditions habitable today. We also think we
know that somehow the ocean's climate buffer crashed, perhaps

because it crossed a crucial tipping point, leaving the planetary climate to cascade into Micawber-like bankruptcy. If we are to discover the well-kept secrets around our own origins, we need to know a little more about what life was like back then. How on earth did life cope with Snowball Earth, and what biological legacy did it leave us?

7

Fossil Records

The Cryogenian ice age represents a pivotal moment in the evo-lution of life. Definitive animal fossils are known only from later rocks, while most earlier life forms were microbes. Rather than being a life-ending calamity, perpetual deep freeze framed the subsequent rise of complex life. Following the great melting, ani-mals radiated in waves during the ensuing Ediacaran Period (635–538 million years ago), opportunistically exploiting newly oxygenated environments wherever they could be found. Investi-gating how life survived Snowball Earth and then flourished into more muscular types that could move, see, excrete, and build shells may provide clues to our own evolutionary origins.

Charles Darwin was already aware in the first edition of the *Origin of Species* that his new theory of evo-lution by natural selection lacked support from the fossil record. This deficit was more than a little em-barrassing, especially seeing that, before modern genetics, fossils were the only firsthand witnesses of evolution at his disposal. In 1859 he wrote:

To the question why we do not find rich fossiliferous deposits belonging to these assumed earliest periods prior to the [Cambrian] system, I can give no satisfactory answer. Nevertheless, the difficulty of assigning any good reason for the absence of vast piles of strata rich in fossils beneath the [Cambrian] is very great.

Unfortunately for Darwin, his famous "dilemma" was not resolved during his lifetime and is still among the thorniest controversies in all of science. Why did so many different types of large, complex, shelly, and energetic creatures, surprisingly modern in their appearance, appear to radiate around the planet so abruptly at the start of the Cambrian Period? Martin Brasier, in his book *Darwin's Lost World*, summarized three possible solutions in rather engaging terms as Lyell's Hunch, Daly's Ploy, and Sollas's Gambit. The first of these refers to Charles Lyell's predilection for gradual change. Lyell, Darwin's friend and mentor, believed that the Precambrian world teemed with life both on land and in the sea but that the more piecemeal, deformed, and cooked nature of the older rock record was to blame for the lack of fossil evidence. However, even in Lyell's day, this was not a very convincing argument, and Darwin admitted as much in another passage of his famous book: "The geological record does not support the idea that the older a formation is, the more it has suffered the extremity of denudation and metamorphism." In other words, if animals existed, we should be able to find fossil evidence.[1]

Reginald Daly also preferred gradual rather than catastrophic change, but instead of blaming the poor rock record, he proposed that Precambrian animals had not left any traces of their existence simply because they had no hard parts that could be fossilized, thus lending the false impression of a "Cambrian explosion" of life. The reason for the sudden appearance of shells, he surmised, was that the oceans were not supersaturated with respect to calcium carbonate before the Cambrian. William Sollas, writing in 1905, already knew that Daly's arguments held no water. After all, Precambrian limestone formations provide abundant proof of the supersaturation of the oceans. Sollas agreed that the explosion was an artifact of preservation, but disagreed with the suggested cause, arguing that the Cambrian explosion marked a biological event when animals first evolved the ability to make shells. These two contrasting but perhaps not entirely mutually exclusive ideas, permissive environment versus biological innovation, still dominate a lot of thinking about the emergence of animal life on earth.

In Darwin's time, the puzzle of the Cambrian explosion was much more acute than today. Today, we know of definite Precambrian animals. For example, we know that cnidaria, the group that includes jellyfish, lived in Ediacaran seas from their goblet-shaped shells. We can even infer worm-like *bilaterian* animals, possibly lophotrochozoans, the group that today includes mollusks, from their characteristic traces. Back in Darwin's day, however, Cambrian trilobites (trilobed arthropods) were the oldest known fossils, and there were no convincing Precambrian examples at all. To the embarrassment of nineteenth-century proponents of evolution, trilobites were evidently armored bugs that would not look entirely out of place today. Faced with such a glaring absence of evidence, Darwin boldly postulated that fossils would eventually be found that would reveal what the ancestors of Cambrian animals looked like during the vast time required for life to evolve from molecular soup into complex, multicellular life. A long Precambrian was in line with the ancient vintage awarded our planet by many other geologists of the day, and indeed ever since James Hutton famously peered into the abysses of time. The notion that life only began to get conspicuous quite late in earth history was enshrined in the name, borrowing from Greek, that was given to its final eon, the Phanerozoic, meaning "visible life." Before we settled on the "Proterozoic," or "early life," for the preceding eon, it was often referred to as the "Cryptozoic," or "hidden life."[2]

Like all good scientific hypotheses, Darwin's idea made testable predictions that we can now verify. Yes, Precambrian time was indeed enormously long. Over 88% of the earth's history had already been and gone by the time trilobites emerged. Another of Darwin's predictions was that animal fossils would be found that predated trilobites, and this claim was also verified, necessitating repeated redefinition of the Cambrian system ever downward. The Cambrian Period is now reckoned to have begun twenty million years earlier than the oldest trilobites and is formally defined as beginning close to the first appearance of both shelly and trace fossils, indicating the evolution of armor and mobility in mostly bilaterally symmetrical animals, or bilaterians, like mollusks and worms. Precambrian precursor animals, and even soft-bodied "bilaterians," are also known in fossil form from the preceding Ediacaran Period. From our anthropocentric viewpoint, this is intriguing because we, like worms, are bilaterians. The abruptness of the Cambrian explosion has been replaced by a long, faint history of evolving complexity. Of the 150 million years from the Cryogenian to the Ordovician Period, only the final quarter contains arthropods, of which the trilobites are the most renowned, as well as other identifiable animal phyla. It would therefore be tempting to suggest that complexity arose only after Snowball Earth, as such near-eternal deep freeze would surely have caused a calamitous extinction. That would overlook, however, the awkward fact that modern genetics predicts that the last common ancestor to all modern animals already existed before the Cryogenian, which means the path taken by life from the Phanerozoic Era right up to our own existence might be contingent on what conditions were like on the snowball. So, just how complex was life before perpetual winter set in?[3]

Two decades after his international field excursion to western Mongolia, Martin Brasier had settled comfortably into the role of a fatherly, white-haired Oxford don, when my wife, the geochemist Ying Zhou, invited him on a trip through a part of her field area: the Proterozoic strata of eastern China between Nanjing and Beijing. Martin had become something of a minor celebrity, having written two books about his life in science, and most controversially a series of scientific papers questioning the biogenicity of the world's oldest fossils, which had until then been three-and-a-half-billion-year-old putative cyanobacteria of canonical status from western Australia. Whatever one makes of those intriguing ribbons, Martin undoubtedly raised the bar by emphasizing the need for a null hypothesis. His work ensured that sensational claims that interpreted suggestive shapes in ancient earthly rocks or Martian meteorites as signs of life would be given short shrift unless stringent criteria could be met. Martin maintained that there was no robust evidence for life in the Archean (yet) and that bacterial life—once it got started—had dominated ecosystems until much later, when life first got big enough to see with the naked eye sometime during the Neoproterozoic Era. Martin was fond of reminding us that science was the measurement of doubt, and now, like a true skeptic, he was daring everyone to put up or shut up.[4]

On our way to Beijing, Ying and I had planned to pick up an elderly retired geologist whom we had worked with before.

After our minibus arrived in Tianjin, however, Shixing Zhu surprised us by inviting us to his home for a "cup of tea." As it turned out, this was only a ruse to show us some fossils that he hoped would satisfy Martin's null hypothesis for ancient complex life. Zhu, now in his eighties, is another of those geologists of an earlier generation whose knowledge of regional geology seems impossibly vast, as he has traveled all over China, forensically mapping and describing rocks both on his own and as part of a geological team. For much of his working life he had been employed in a regional office of the Chinese Academy of Geological Sciences. Finally retired, he could now afford the time to read the international literature more closely and had been puzzled to discover that scientists had recently gotten excited by some Ediacaran fossils, called the Lantian biota, simply because they were large. As we sipped green tea in his living room, surrounded by box upon box of far older fossils, he told us how his reading had inspired him to share some of the extraordinary finds he had made over the years.

Zhu's incredible fossil collection was not entirely unknown, as he had initially brought it to the attention of the global paleontological community via the pages of the journal *Science* back in 1995. In that paper he described leaf-like fronds up to several centimeters in length that, incredibly, were more than 1,600 million years old, more than three times older than the oldest trilobites. Centimeter-sized leaf litter does not sound like much to write home about, but up until then the only signs of large organisms in rocks that old were simple spirals and chains of beads that could merely have been large bacteria, or perhaps not even fossils at all. Except for these tiny smudges, everything else this old was microscopic. By contrast, Zhu's fossils had intriguing features that could potentially place them within the eukaryotes, the same group as pretty much all large things that exist today (animals, plants, fungi, and algae). What makes eukaryotes different from other organisms is that they have more complex cells, which include a nucleus and organelles, some of which once lived independently as bacteria. Although suggestive, the Chinese fossils were still too simple to give conclusive evidence of a eukaryotic origin, and the paper was widely overlooked. What the geological community did not know, however, was that Zhu's work did not stop in 1995 and that he had, during the succeeding two decades, amassed an impressive collection of much larger and more ornate organic-walled fossils.[5]

That afternoon in his living room, Zhu was boyishly keen to show off his personal fossil trove, to Martin's undisguised amazement. Many Precambrian fossils are tiny, uninspiring tapes and balls, but these new fossils left little to the imagination, looking for all the world like folded palm leaves that had been whipped up by a storm and dumped on the seafloor just yesterday. The new fossils were from a slightly younger rock body, but still 1.56 billion years old, called the Gaoyuzhuang Formation after a place in the Jixian national park, near Tianjin. Most Gaoyuzhuang fossils are only fragments, but some are nearly complete and, for their time, preposterously huge, reaching up to thirty centimeters in length and eight centimeters in width. Some are circular disc-like impressions, but most are striated fronds that were likely rooted to the seafloor by holdfasts, suggesting a degree of cellular specialization. In other words, not only were these organisms multicellular, comprising hundreds of individual cells, but they were not simple colonies of identical cells. Some cells had their own separate purposes within the overall organism, just like trees have roots and leaves and stems. With the help of expert colleagues in the United States and China, Zhu eventually published his findings in 2016 to widespread astonishment.[6]

The spectacular finds from North China represent the

world's oldest convincing eukaryote fossils. They lived by photosynthetically harnessing the sun's energy while anchored on the seafloor, like seaweed today. In the two decades between Zhu's two papers, smaller, microscopic organic-walled fossils called acritarchs had also been discovered in the same suite of rocks. The intricate ornamentations of the acritarchs strongly implied that these also were eukaryotes that had lived a planktonic existence, at least during one phase of their lives. Intriguingly, the Gaoyuzhuang fronds appear to have lived in oxygenated waters, and follow a major carbon isotope excursion, just like the ones that characterize the Neoproterozoic oceans of almost a billion years later. It is just too tempting not to suggest that these unusually large and complex fossils were able to emerge because the shallow oceans had temporarily gained some oxygen and lost their turbidity, allowing the power of the sun's rays to extend deeper into the Proterozoic marine environment. A pattern is starting to emerge that links changes in life with changes in its environment, and we will return to this recurring theme later on. But let's not get ahead of ourselves. We still know little about the world of the early Proterozoic, but what we do know suggests that, as Darwin had predicted, cryptic evolutionary steps were indeed taking place, and with time we will know much more about their nature and importance. Once our group was back on the minibus, animated conversations ensued on the way to Zhu's still-secret outcrops, where we would be able to unearth the same fossils from their rocky tombs. Martin had coined the term "boring billion" some years previously, but it now seemed an appropriate time to abandon the epithet. Clearly, evolution was making great strides, albeit only episodically and in rare ecological niches.[7]

Despite such early promise, there is no sign that these biological innovations ever dominated global primary production before the Cryogenian Period. Indeed, assuming that size is meaningful in a biological sense, it is intriguing that life did not get this large again for another billion years. Although fossil eukaryotes have been found throughout the boring billion, they were so unassuming that they did not even leave a molecular fingerprint. Eukaryotes are characterized by cell membranes made of steroids, which are four-ringed organic compounds that alter to steranes after death and burial. However, no indigenous steranes have ever been detected in rocks older than about 800 million years, after which steranes appear in abundance worldwide, and some in forms with no modern counterparts. You may recall that the first Snowball event, the Sturtian ice age, began about 717 million years ago, tantalizingly close by geological standards. By contrast, hopanes, which are bacterial biomarkers, dominate the organic signatures of pre-Cryogenian rocks. Yet the Gaoyuzhuang fossils suggest a more nuanced story. Surely, there were at least episodes when eukaryotic life ran riot, even if only for a short time? How can we reconcile such apparently contradictory evidence? It could be, against all other available evidence, that these were not, in fact, eukaryotes. Or, perhaps, ancient eukaryotes did not produce steroids. Or maybe all the molecular evidence has been digested in the microbial carpets that draped the Precambrian seafloor. Only time will tell which of these, if any, is correct. But for now, at least, all the evidence suggests a relatively minor role for algae before the Neoproterozoic Era.[8]

The case for early eukaryotes gets stronger as we draw closer to the Cryogenian ice ages. The oldest definitive fossil red algae, for example, were discovered in Canadian rocks aged about 1,050 million years old. They were reported more than thirty years ago by Nick Butterfield, then a student of the renowned paleontologist Andrew Knoll of Harvard University. Andy Knoll and his students have dominated Precambrian paleontology ever since, and to such an extent that one could learn almost everything worth knowing about the early fossil record through their work alone. For his myriad personal discoveries, pioneering isotopic studies, and inspirational mentoring of students, Andy was awarded the Crafoord Prize in 2022, widely portrayed as the Nobel Prize of the geosciences. Like so many of Andy's past students, Nick Butterfield is now himself a professor at the other Cambridge, in England. An entertaining speaker and all-around maverick, he will appear again later. Nick's red algae display cellular specialization and, thirty years on, are still the oldest fossils, except perhaps for cyanobacteria, that can be confidently assigned a specific taxon. Much more recently, fossilized green algae were discovered in rocks of similar age from China. Macroscopic, multicellular, and clearly exhibiting cellular specialization, both of these examples differ from the Gaoyuzhuang fossils in that, although much smaller, they can be readily placed into a modern algal group. It is currently unclear whether these rare finds are tantalizing glimpses of cryptic algal biodiversity throughout the boring billion, or whether they mark genuine evolutionary diversification toward its end.[9]

Fast forward a decade and another student of Andy Knoll, Susannah Porter, published an equally astonishing discovery of Precambrian eukaryotes in the form of fossilized amoebae from the Grand Canyon. Subsequent work by Susannah, now Professor Porter, has shown that these and other fossil "protists" can be found all over the world, but always in rocks younger than about 800 million years. You may recall that this time window coincides perfectly with the first occurrence of eukaryotic sterane biomarkers, but these, strangely, are dominated by only one type, cholestane, and may also have formed from protistan cholesterol. The very first signs of biomineralization also appear around the same time in the form of tiny phosphatic scales that once surrounded single-celled green algae.

These were first discovered by Phoebe Cohen and, yes, you guessed it, her PhD supervisor, Andy Knoll. Importantly, amoebae do not use sunlight like algae but instead are predatory single-celled protists that consume other single-celled organisms. It is tempting to speculate that scales, and biomineralization in general, evolved to protect prey from attack, and suspicious holes have indeed been found in the brain-shaped acritarch *Cerebrosphaera buickii* as well as in testate amoebae of this time. Was there a cryptic Tonian "arms race," foreshadowing the much later Cambrian warfare that led to the evolution not only of shells, but of diverse forms of weaponry and protection in equal measure?[10]

The "heterotrophic" metabolism of these early single-celled protists is similar, and seemingly ancestral to our own omnivorous ways. But are amoebae or other protists true animals? And if not, what distinguishes them from simple animals? Not a great deal. The most primitive creatures that are still considered to be animals are probably the sponges, the sister group to all other animals. Although sponges are multicellular and hence referred to as bona fide "metazoans," another word for animals, they are essentially colonies of cells that can morph into other types and even move around, with the result that sponges have no specific body shape. Sponges, moreover, have no nerves, guts, or blood but subsist on the food and oxygen that gets pumped through their bodies by the tiny whips of thousands of choanocyte cells, which line the cavity walls. It was Felix Dujardin, a French naturalist, who first noted that protists were not simply miniature versions of known animals but instead formed their own separate group. In 1841, Dujardin observed a similarity between choanocyte cells and a group of single-celled creatures called the choanoflagellates and proposed, not unreasonably, that the choanoflagellates are the closest living relative to all extant animals. Although the details are

foggy, DNA studies have added weight to his theory, implying that the leap from a single-celled protist to a stem group
metazoan, a bit like a primitive sponge, perhaps, and the last
common ancestor of all later metazoans, is not inconceivable.
Dujardin was a contemporary of Darwin, but at long last we are
getting closer to finding out when that crucial step to multicellularity occurred.[11]

Despite many false alarms and a lot of fruitless searching, it is still the case that there are no convincing physical
signs of animals in any pre-Cryogenian or Tonian rock succession. Many modern sponges build a robust framework out of
tiny needles of silica glass, called spicules, but characteristic
spicules have only been found in Cambrian and younger rocks.
Indeed, multicellular organisms, other than bacterial colonies,
are rare to absent in the Tonian as well as the subsequent Cryogenian Period. Frustratingly, as the world descended into its
long, cold winter, fossil records of all types went eerily silent.
With very few exceptions, only organic-walled spheres of uncertain affinity are known from the entire Cryogenian Period.
Organic geochemists once proposed the existence of demo-
sponges, but, although molecular clocks indicate that sponges
had evolved by then, enthusiasm for that particular biomarker
interpretation is waning fast. The fossil record for the entire
interval between about 717 million years ago and 635 million
years ago is not just controversial, it is virtually non-existent.
Yet we know that various groups of many forms of bacteria,
red algae, green algae, protists, and specifically our ancestors
survived somewhere, somehow. We know this not only because of morphological similarities between fossils and extant
organisms but also from genetics.[12]

DNA helps us to reconstruct the tree of life by comparing in minute detail the mutations that have collected over time,
allowing us to work out the sequence of diversification events

from a common ancestor. Biologists can also work out the timing of those events by using the known ages of groups that have
a good fossil record. Attempts to work out when the last common ancestor of all living animals was alive have been made
repeatedly over the years with startlingly different results and
implausibly high levels of precision. The most recent attempts
have applied sophisticated statistics, called Bayesian inference,
to arrive at a more parsimonious picture. These highly specialized studies were the work of a group of mathematically
minded biologists from University College London and the
University of Bristol led by Ziheng Yang and Phil Donoghue,
respectively. Amazingly, the first trials of the new methods
suggest that major diversification events occurred *during* the
Cryogenian Period on Snowball Earth.[13]

The most recent genomic updates from the Bristol group
broadly confirm this picture but under even tighter constraints,
placing the last common ancestor of all animals close to the
end of the preceding Tonian Period (Figure 14). If this new reading of the genetic code is correct, then not only sponges but
also other ancestors of modern phyla persisted through both
the Sturtian and Marinoan ice ages. The last common ancestor
of all animals higher than sponges, the eumetazoans, seems to
have emerged on the Cryogenian snowball, while bilaterians—
from the same analyses—arose toward their end or possibly
shortly thereafter. Once considered an impassable dead end
for all higher life forms, Snowball Earth has strangely become
the crucible of biological modernity. Yet we have absolutely no
fossil evidence in support of such a claim, other than a lack of
animal fossils before the Cryogenian Period and a plethora of
animal fossils afterward, as we shall find out below. But how is
it even possible for complex life to survive such extreme cold,
let alone proliferate and diversify? Until recently we had no viable answer to this question, but studies of life on today's frozen

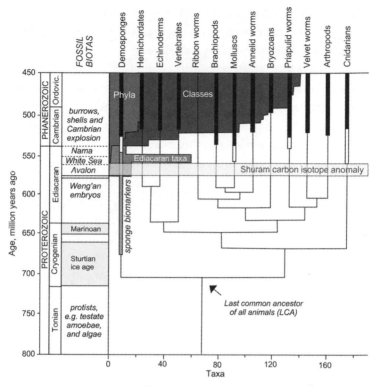

Figure 14. Early animal evolutionary history. Reconstruction of life's evolutionary steps during the emergence of animal life showing the accepted fossil record of key animal groups (black bars), suggestive fossil record (open bars), and speciation points back in time based on differences in the DNA of modern phyla. The animal evolutionary record shown here is based on an article from 2011 (Erwin, D. H., Laflamme, M., Tweedt, S. M., Sperling, E. A., Pisani, D., Peterson, K. J., "The Cambrian conundrum: Early divergence and later ecological success in the early history of animals," *Science* 334 (2011): 1091–1097), but recent updates suggest that the last common ancestor of all animals may have lived no earlier than the Cryogenian Period. Vertical (age) error bars on branching points are not shown for clarity.

ice caps have led to a rethink. It turns out that a globally frigid climate might have been just what the doctor ordered, at least as far as biodiversification is concerned.

If, like me, you grew up on a diet of nature programs showing nothing more than penguins and seals on the barely habitable margins of Antarctica, you might indeed be forgiven for thinking it a barren place. Once the Drake Passage opened between South America and the great white continent, the circumpolar current effectively removed Antarctica from warmer, tropical waters forever. Antarctica then cooled inexorably over the last fifty million years, during which time even gradual adaptation to the extreme climate became impossible for all but a few hardy species. And they really only survive by clinging on at the edges of the ice caps to fish the thawing sea each long summer. However, this is not an entirely true impression. In Antarctica, as also in Greenland, a diverse range of eukaryote species can be found alive, but trapped in myriad holes, hermetically protected from the inclement world above by a thin ice ceiling. These tiny ecosystems originate when windblown dust, called cryoconite, melts the surrounding ice, forming bubbles of liquid water. The relative darkness of the dust, compared with the surrounding white ice, changes the albedo at the surface, leading eventually to tiny ponds where bacteria, algae, and, amazingly, even animals can thrive.

Cryoconite dwellers are by necessity the hardiest of critters. Rotifers, for example, are tiny, free-swimming animals, some of whom do without sexual reproduction. Some rotifers even do without males entirely. Females produce eggs so robust that they can lay dormant for decades, their embryos protected from drying out by a three-layered casing. But it is not just rotifer eggs that are tough. Even some adults can remain in a state of total desiccation for years at a time; a handy trick in challenging environments. When not pretending to be a shriveled corpse, rotifers feed on detritus, like dead bacteria, and would, you might think, be at the top of the food chain in cryoconite holes were it not for the existence of yet even sturdier sorts. Tardigrades, or water bears, are possibly the most resilient of all known animals. Their endurance records include surviving the most extreme levels of temperature, pressure, radiation, starvation, dehydration, and air deprivation as well as desiccation. They have even been shown to endure the rigors of space for ten days, having survived being put into orbit around the earth on the *outside* of the rocket.

It is easy to imagine how cryoconite holes would have created innumerable miniature ecosystems on the surface of the snowball. There would have been no shortage of airborne dust, which would have collected over time as a result of wind erosion, volcanism, and the occasional meteorite shower. The increasingly darkened surface might even have helped to nudge the earth closer to escaping its perpetual winter due to the effects on planetary albedo. Demonstrating this, however, may prove difficult, as all the evidence has literally melted away. Nevertheless, it is an intriguing thought that myriad, isolated populations of single-celled eukaryotes could have formed the ideal breeding ground for biological innovations in an extreme form of what biologists call "allopatric speciation." And there may be other reasons for thinking that Snowball Earth favored the evolution of complex multicellularity given the intrinsically altruistic nature of animals. You may not regard animals as altruistic, but in a biological sense they can be described as such because their cells have abandoned any independence, working cooperatively toward the common goal of passing on genes. Their altruism shows itself not only in cellular differentiation but also in apoptosis, which is the process by which cells are pre-programmed to commit suicide when there is danger for the whole organism. One of the reasons why cancer is so

feared is that cancer cells have become anomalously independent. Almost like microorganisms, they move around freely,
metabolize differently from all other cells, and do not practice
cell death. One of the most interesting hypotheses to explain
why animals evolved when they did postulates that the sheer
hostility of conditions on Snowball Earth forced organisms
into cooperative partnerships, not only in a symbiotic sense
as with lichen, which are collaborations of mutual benefit between different kingdoms of life, but also within organisms,
whose survival benefited from existing as colonies of increasingly specialized cells.[14]

Further speculation on this topic lies outside the scope of
this book, but it is important to recall that Darwin's Dilemma
and much subsequent head-scratching derived from the relative abruptness of animals' appearance on earth. Recent fossil discoveries, DNA advances, and improved geochronology
change all that. The first Snowball episode lasted, incredibly,
for more than fifty-seven million years, enough time even for
continental ice sheets to sublimate, or slide, entirely onto the
frozen oceans. Nevertheless, plausible protistan ancestors of
all modern animals appear to have survived to diversify into
all modern phyla within the following 200 million years. Having sufficient time for animal evolution seems no longer to be
an issue.

Diversification could have been followed by radiations
after melting, and this seems likely to have occurred during the
relatively short non-glacial interval between the Cryogenian
ice ages. Biological novelties, spawned during the immensely
long Sturtian ice age, would have been able to spread throughout the now liquid oceans, where they could finally meet and
compete with other new forms. Although evidence for this initial radiation has been slow to appear, recent discoveries, most
specifically the appearance of enigmatic discoidal fossils, called

Aspidella, in similar deep marine environments in which they would make a second appearance during the later Ediacaran radiations, would seem to confirm the genetic evidence for diversification. The non-glacial interval is also the first time that eukaryotic sterane biomarkers appear in abundance and diversity, suggesting that this was a time when photosynthetic algae began to compete more effectively with the ancient cyanobacteria. We know so far of only one instance, however, of large-rooted fronds during the Cryogenian Period, and these were also found by our old friend Shixing Zhu of Tianjin. Much more remains to be discovered, but for the time being the Cryogenian fossil record is keeping its cards close to its chest.[15]

As we have already seen, the Marinoan event, by comparison, was much shorter and its ending more abrupt. Huge expanses of frozen tundra were flooded by the sea as vast ice sheets melted, raising the sea level by hundreds of meters. In a geological instant, a second new set of glacial ecosystems, having evolved independently for millions of years, met in the rapidly warming post-glacial ocean. Out of the ice and into the fire came new biological innovations that had evolved to endure extreme hardships: not just freezing and desiccation, but something equally damaging, exposure to oxygen, due to the greater solubility of gases in cold pools of fresh water beneath an oxygenated atmosphere. But now these novel metazoans were suddenly and without warning exposed to warm or even hot, less oxic, salty water. This sudden changing of the rulebook likely led to extinctions and radiations in equal measure. If Snowball Earth had turned out to be a "good thing" for biodiversity, then perhaps the abruptness of the return to what we might regard as normality was the genuine catastrophe.

Until quite recently, very little was known about the organisms that lived on earth after the Cryogenian glaciations. Indeed, there was a huge fifty-million-year gap in the fossil record from the cap dolostones until large animals finally arrived on the scene during the second half of the Ediacaran Period. Rocks do not give up their secrets easily, and despite years of dissolving them to reveal their organic contents, scientists had not found anything very exciting in the classic lower Ediacaran succession of South Australia (or indeed anywhere else). The absence, it turned out, was not always real, but an artifact of the deep weathering that is typical of the scorched red continent. As soon as unweathered drill cores were investigated in the 1980s and 1990s, a very different picture began to emerge. Rather than being barren, many lower Ediacaran rocks quickly revealed that they had originally been full of organic-walled microfossils, or acritarchs, that were unlike anything that had been found before. Many of these fossils were found and described by the expert eyes of Kath Grey of the Western Australian Geological Survey. These "large spiny acritarchs," or LSAs, have subsequently been found all over the world and dubbed the Pertatataka Assemblage, after the formation in Australia where they were first found. The very oldest are found in China, just above the basal Ediacaran cap dolostone there, which has been dated to 631 million years ago. Such LSAs have been compared favorably by some authorities with the resting cysts of modern animals.[16]

We will find out in chapter 9 what some of these acritarchs might really have been, but for now it suffices to say that we are still not entirely sure. The clue is in the name. *Acritarch* is the bucket term we give to organic-walled fossils of uncertain taxonomic affinity. Strangely, the large spiny acritarchs departed as abruptly as they arrived, and in this case the disappearing act does not appear to be an artifact of weathering. No LSAs are found anywhere on earth after the "Gaskiers" glaciation circa 582 million years ago, but that's not for lack of trying. In spite of some microscopic clues provided by the LSAs, the first half of the Ediacaran Period continues to withhold most of its biological secrets, although tantalizing glimpses, such as the unique LSAs, give us confidence that answers will be discovered. The second half of the Ediacaran Period, by comparison, is a fossil hunter's paradise that prepares us for the riot of new forms and traits that arrive on the scene during the subsequent Cambrian radiations.

We met *Aspidella*, an enigmatic disc-like fossil, once already in this chapter, as it appears briefly before the Marinoan ice age in northwestern Canada. It reappears in the deep marine sedimentary rocks in Newfoundland that sit above the approximately 582-million-year-old Gaskiers diamictite there. *Aspidella* was first discovered in 1868 by the Scottish geologist Alexander Murray and was the first bona fide Precambrian fossil ever to be described as such. After the great American paleontologist Charles Doolittle Walcott, who does not seem to have viewed the fossils himself, poured cold water on the discovery, few people took any more notice of *Aspidella* until about a century later. It is an unfortunate fact that disc-like features are unremarkably common in sedimentary rocks and many of them are not biological at all, hence the importance of the null hypothesis. The *Aspidella* discs finally grabbed attention following the discovery of more obviously biogenic fossils from rocks of a similar age in Newfoundland and elsewhere.[17]

Geologically speaking, Newfoundland is the twin of the British Isles. Both of them are divided through the middle by a suture zone that marks the line along which the continent of Avalonia (England, Wales, southern Ireland, and southern Newfoundland) collided with Laurentia (Scotland, northern Ireland, and northern Newfoundland) in the Paleozoic Era. Until the growth of the Atlantic Ocean rudely tore them apart, parts of Newfoundland had much more in common with England than either of them did with Scotland, so it is no surprise, in hindsight, that it was in Charnwood Forest in Leicestershire, England, where, in very similar rocks, fossils of the same type and age were found in 1956 by a local fifteen-year-old schoolgirl named Tina Negus, and then rediscovered a year later by the sixteen-year-old schoolboy Roger Mason. Roger later made a career as a geologist—although, ironically, he eschewed sedimentary rocks for their igneous counterparts—becoming renowned as a most enthusiastically animated lecturer at UCL before, and Wuhan University after, his so-called "retirement." The fossils are not particularly easy to appreciate with the naked eye, having very low relief, and can best be viewed by the low-angle lighting of dawn or twilight. Mason had the impressive foresight to take a rubbing so that he could show his father, who then showed it to a local geologist, Trevor Ford. Ford named the fossil *Charnia masoni,* and the rest, as they say, is history. The Charnwood fossils surpass even the sizes of the largest Gaoyuzhuang fronds of northern China, breaking a record set an incredible billion years earlier.[18]

Charnia has now been found all over the world, but we still don't know how it made a living. It clearly could not move and so needed its food to come to it. However, it lived beneath the photic zone on the deep seafloor, so it could not have harnessed the energy of the sun like a plant or a symbiotic animal or lichen. The consensus today resides with a metazoan interpretation, in which case it would be the oldest definitive animal fossil known. These now iconic, quilted fronds, whatever they were, and however they lived, were the vindication that Darwin had sought a century earlier. In *Attenborough's Journey,* the great naturalist and filmmaker David Attenborough expresses his profound regret that he had not given those rocks his full attention. Attenborough had attended the same school as Mason years earlier but, knowing the great age of the Charnwood rocks, had decided to do his childhood fossil hunting elsewhere.

These days, England and Newfoundland give up a little more of their fossil secrets every year. During the heady days after Paul Hoffman's revival of the Snowball Earth hypothesis, a group of paleontologists, led by Guy Narbonne from Kingston University in Canada and Jim Gehling from the South Australian Museum, guided an international group of geologists to the Ediacaran fronds of Mistaken Point, so named because sailors would mistake it in poor weather for a neighboring safe harbor. For many of us, including me, this was our first opportunity to see Ediacaran fossils, and with various new papers hot off the press, supporting or rebutting the idea of global glaciation, the excitement was palpable (Figure 15). Many of the evenings were spent listening to the giants of the field, like Bruce Runnegar and Dolf Seilacher, vigorously debating Snowball Earth and early life. The science writer Gabrielle Walker records some of these exchanges in her book *Snowball Earth.* It was bizarre but strangely appropriate to see scientists from around the world pay homage to these most sacred of fossil sites by treading the treacherous rocky cliff faces in socks so as to leave them just as pristine as they found them.

Thanks to the work of many groups of paleontologists, we know a lot more about the Ediacaran "Avalonian" fossil assemblage. We have discovered that many forms, and even *Charnia,* were originally rooted to the seafloor by disc-shaped holdfasts, from which they swayed with the currents and occasionally got flattened by turbidite flows or the fallout from volcanic eruptions. It seems likely that *Aspidella* was also just such a holdfast, for which we have yet to find a stem. The Newfoundland ash beds are a godsend for paleontologists, as the ash helped to preserve animals' soft parts, far better even than at Pompeii,

Figure 15. Representative late Ediacaran fossil taxa, c. 565–545 million years old, counterclockwise from top right: the classic Ediacaran fossil *Dickinsonia* from Australia is an enigmatic taxon that had an anterior and posterior end but no obvious head or anus; *Rangea* from Namibia (upper left) is one of many Ediacaran "rangeomorphs" composed of leaf-like "frond" elements that show fractal self-similarity on four scales; *Pteridinium* from Namibia is a typical erniettomorph, a group characterized by airbed-like tubes and unusual "glide" symmetry; the first signs of locomotion are found around this time, exemplified here by a trace fossil from Newfoundland that shows the trail of an organism, which had grazed the surface of a microbial mat. None of these fossils can unambiguously be placed into modern animal groups, but all likely relate to early forms of animal-grade organisms. White scale bars are 10 millimeters in length. (Photos Alex Liu)

while at the same time providing the means to date their ages precisely. The oldest of the Avalonian assemblages emerged shortly after the Gaskiers ice age by about 575 million years ago, give or take a few million years, and the youngest no earlier than about 560 million years ago. Some groups, including *Charnia,* lived on elsewhere in the world, although mostly in shallower marine realms. Intriguingly, this time interval matches the late Ediacaran carbon isotope anomaly, which, if you remember from the previous chapter, was the greatest negative carbon isotope excursion of the entire geological record. This "Shuram" anomaly will help us to shed light on some of the secrets of their success, but such anomalies also connect these extraordinary biological radiations with the earlier climatic extremes of the Cryogenian Period.[19]

Paleontologists from around the world still visit the stunning bedding planes in Newfoundland every year. Other treasured secrets, now finally revealed, include huge fronds up to two meters long. Even *Charnia* is now known to reach sixty-six centimeters in length. The first signs of purposeful movement anywhere in the rock record can also be found in these rocks, in the form of surface trails and tiny vertical burrows, possibly made by cnidarians, the group that today counts coral polyps and jellyfish among their number. Many of these fossils were discovered by students of Martin Brasier. On a recent visit I made to the Bonavista peninsula in Newfoundland, some of them fondly recalled how he had visited them one day in 2008 to catch up on the day's work. They showed him their most beautiful fossils, but something else caught his eye. "Yes, those are great," he said, "but what on earth is that?" he asked, pointing to an impression like a squashed teabag. Quite by chance, and with the help of the twilight shadows, he had stumbled across what is still the oldest example of fossilized muscle tissue. Although Martin preferred the more prosaic *Melrosia,* after the nearest town of Melrose, that name had already been taken, so he caved to Alex Liu's suggestion to call the new fossil *Haootia,* which means "demon" in the native Beothuk language. The most parsimonious explanation is that Martin's demon was also some sort of early cnidarian, possibly the oldest fossil that can be assigned a modern metazoan phylum.[20]

Avalonian macrobiota left two- and even three-dimensional molds and casts of organisms that had no hard parts whatsoever. The soft parts, skin, and muscles of Ediacaran metazoans are, somewhat unusually for fossils, laid bare. Some later forms, like the Ernettiomorphs, seem to have been like bloated quilts arranged as rows of simple inflated tubes filled with some fluid. Others, like *Charnia,* have been assigned to the rangeomorphs, which had a unique body plan among animals. Intriguingly, rangeomorphs were composed of a complex series of self-similar, or fractal, repeated units. Earlier workers had originally shoehorned fossils such as *Charnia* into modern animal phyla like cnidaria or thought they might be algae or fungi, but their unusual fractal symmetry makes such propositions unlikely. Dolf Seilacher proposed that such quilted Ediacarans formed their own extinct kingdom, the "Vendobionts," although most workers feel that they still fit best as animals of some sort, albeit perhaps a sister group to most modern metazoans. Either of these ideas implies that they represented an extinct evolutionary line or failed experiment on the path to the modern animal kingdom. However, is this really fair? Would it be fair to consider the dinosaurs to have been failed experiments just because, with the exception of birds, they all went extinct? After all, they lived for far longer than humans have been around. Even though they appear to have suffered at least partial demise before the Cambrian explosion, the Avalonian biota must have been excellently suited to their ambient environments to have survived for so long. If so, then what made fractal geometry a success for the final thirty million years of the Ediacaran Period?

Some intriguing work has been done on the morphology of Ediacaran rangeomorphs. Despite looking like a rugby ball, *Charnia* had a small, bulbous holdfast that anchored it to the seafloor, a short stem and a much longer body, comprising rows of tightly stacked first-order branches, up to twenty-five second-order branches, and smaller third- and even fourth-order branches in larger specimens. Some exceptionally well-preserved examples have revealed exquisite 3D detail of the serial quilting, exemplified by a tiny rangeomorph discovered in Spaniard's Bay, Newfoundland, by Guy Narbonne. Mathematicians have shown that all rangeomorphs maximized their surface area to an extent not seen in any extant animals, but reminiscent of internal organs, such as lungs, stomachs, and intestines. Could the functions of our internal organs be analogous to Ediacaran rangeomorphs? A growing chorus of paleontologists seem to think so.[21]

Internal organs, like lungs or guts, commonly show intricately folded surfaces in order to maximize their contact with the oxygen or food passing through those organs. The direct absorption of nutrients through the skin is called osmotrophy, and Marc Laflamme, erstwhile student of Guy Narbonne, has championed this feeding strategy for all modular Ediacarans, including the curiously fractal rangeomorphs. Marc, now a professor in Toronto, together with two colleagues, Shuhai Xiao and Michal Kowalewski, noted the absence of features that would have supported alternative lifestyles, and were left with osmotrophy as the sole remaining option. According to Sherlock Holmes, if all avenues have been exhausted, then the remaining possibility, however improbable, must be true. However, there is a problem. Although many animals supplement their nutrient intake with opportunistic osmotrophy, no animals larger than bacteria use this as their sole means of making a living. There just isn't enough food around to make it worthwhile. Marc and others have estimated that the exceptionally high surface-to-volume ratio of many Ediacarans was indeed similar to modern osmotrophic bacteria, which lends support to the osmotrophic hypothesis. However, for this lifestyle to support such large animals, the levels of assimilable nutrients in their ambient environment must have been extremely high, potentially orders of magnitude higher than in today's oceans. High amounts of organic food in the Proterozoic oceans are consistent with the carbon isotope evidence discussed in the previous chapter, but interpreting those pesky negative carbon isotope anomalies has proved to be highly contentious, as we will see later.[22]

We have come such a long way from Darwin's Dilemma in such a short time that it can be hard to keep up with all the new findings. Geochronology has improved so much that we can now place each new fossil into a global evolutionary framework. The Precambrian biosphere that has emerged is a very different place from today's, not only in terms of the life forms that were present, but more importantly in terms of the lifestyles that were viable. Until late Tonian times, eukaryotic red (or green) algae could make a living, but only at the peripheries. Furthermore, heterotrophic lifestyles among complex eukaryotes were all but absent in a world dominated by microbes. But all this seems to have changed in the run-up to the Cryogenian glaciations, which saw a cryptic diversification that eventually led to all metazoan life to come. Although some forms of life that might be familiar to us today appeared after the Gaskiers glaciation on the ancient ocean margins of Avalonia in Newfoundland and England, most of the seafloor was populated by strange, quilted pillows, anchored immobile to the seafloor, basking in a nutrient- and organic-matter-rich stew with no modern counterpart, except perhaps for raised peat bogs. Very little moved on the Avalonian seafloor, except as a result of currents or storms. Although Darwin's Dilemma places emphasis on the rise of modern animal phyla and their behavioral traits, such as building armored shells or digging burrows, those lifestyles were probably not viable in the environments in which the very earliest metazoans evolved. However, things were about to change. And key to that change—that would eventually lead to the Cambrian "explosion"—was oxygen.

8

Oxygen Rise

The surface of our planet has become increasingly oxidized. This inexorable trajectory led to today's atmosphere and oceans replete with free oxygen. Geochemical tracers tell us how and when first atmospheric oxygenation (Great Oxidation Event ~2,300 million years ago) and then ocean oxygenation (Neoproterozoic Oxidation Event ~570 million years ago) occurred. Biological radiations of early animals are marked not only by the spread of oxygenated environments, but also by extreme volatility in the extent of ocean anoxia. Life plays an important role in regulating oxygen distribution, but global changes to the earth's oxygen inventory are also subject to grand tectonic events far beyond life's control.

When life originated, our planet must have been entirely devoid of free oxygen. This curious claim was first made in 1924 by Alexander Oparin and in 1929 by J. B. S. Haldane. They both argued that the building blocks of life could not have arisen spontaneously in the presence of such a reactive element. Oparin and Haldane were following directly in the footsteps of Darwin, who had postulated that inorganic molecules had sparked into life in some "warm little pond," helped on their way by ammonia, phosphorus salts, heat, light, and lightning. Stanley Miller and Harold Urey reproduced Darwin's "primordial soup" in a classic experiment in 1952, in which simple hydrocarbons were synthesized in the absence of oxygen, accompanied by electrical charges to simulate lightning strikes. Darwin's warm little pond has since become a hot tub of debate as scientists like Nick Lane at UCL have begun to question whether a well-stirred "soup" could ever provide enough chemical disequilibrium of the kind that energizes all living cells today. White smokers, which are chimneys on the seafloor where alkaline fluids come into contact with relatively more acidic seawater, seem now to be a more suitable crucible for the origin of life. Nevertheless, the notion that life originated in the complete absence of free oxygen remains unchallenged and has interesting implications when we think about the evolution of complex life.[1]

Given that the early atmosphere, and therefore also early oceans, was originally anoxic, there must have been a time when oxygen rose to the levels needed to power complex and energetic life forms such as ourselves. In 1959, John Nursall addressed this point in the journal *Nature* in his answer to the question: "Why are there no fossil records of animal life from earlier periods than the Cambrian, the lower reaches of which are characterized by trilobites?" It is illuminating to read that only sixty years ago, the enigma of the Cambrian explosion could be expressed in precisely the same terms as Darwin's, fully a century before. You may recall that 1959 was the year following the first report of Precambrian animal fossils from

England, but Dr. Nursall, a Canadian zoologist, can be forgiven for not having read the *Proceedings of the Yorkshire Geological Society.* Nursall's own solution to the conundrum was
that animals arose as a consequence of an increase in oxygen
levels beyond a viability threshold for aerobic metabolisms.
Thus began a tussle between pro-oxygen and anti-oxygen camps
that continues today, although the framing of the question has
become ever more nuanced.[2]

Of all the necessities of life, we probably take oxygen most
for granted. When describing something easy, we say that it is
just like breathing. But if we were to journey back to before
there were any animals, for instance, I doubt we would find
breathing quite so effortless. We know that free oxygen first
became abundant after the GOE almost two billion years before the Cambrian Period, but if oxygen levels were still low
before the Ediacaran Period, perhaps a further increase in available oxygen could indeed explain the rise of animals that followed the end of Snowball Earth 635 million years ago. To find
out whether oxygen might indeed hold the key to animal evolution, we need a good understanding of what controls the earth's
oxygen budget. As with carbon dioxide, oxygen levels are the
result of a balancing act between sources and sinks, or in chemical terms between oxidant (electron acceptor) and reductant
(electron donor) flux. Considering its highly reactive nature,
it is doubtful there would be much, if any, free oxygen on our
planet in the absence of life. Indeed, life must continuously produce oxygen in order for its production to match its consumption by natural processes such as (oxidative) weathering and
volcanism. The oxygen we breathe is therefore a consequence
not only of its generation through photosynthesis but also of
a critical balancing act between geological phenomena such
as sedimentation, volcanism, metamorphism, and erosion. To
grasp the enormous scale and potential significance of such

episodic processes for the earth's oxygen budget, it might be helpful for us to go on a short field trip.

Every year, I take undergraduate students on a day trip to Tedbury Camp Quarry in a sleepy corner of rural Somerset in southwest England. The trip runs on consecutive weekends, with the first around Halloween and the second around Bonfire Night, which recalls the day in 1605 when Guy Fawkes narrowly failed to blow up the Houses of Parliament in London. If we are lucky, at the end of a hard day's fieldwork, we get to see fireworks and bonfires sprouting up all the way from Stonehenge to London as we journey back on a dark November evening. We visit the quarry to see the best example of the rock cycle within a day's return journey from central London. The quarry floor at Tedbury Camp, once the site of an Iron Age hill fort, separates angled layers of gray limestone beneath from horizontal layers of yellow limestone above. Both sets of rocks contain abundant fossils that reveal the enormous difference in their ages. The lower layers contain numerous finger corals as well as rarer sea lilies (crinoids) and brachiopod shells that populated the tropical reefs of the Carboniferous Period. By contrast, the upper layers contain oyster shells known to have lived 170 million years later during the Jurassic Period. The flat quarry floor that separates them was once a wave cut platform, planed by time and the relentless pounding of the sea. Tedbury Camp reminds students that the visible "rock record" at any one place is missing huge chunks of time, and that reconstructing earth history from such incomplete jigsaw puzzles is a non-trivial task.

It was James Hutton who first recognized that such "angular unconformities" represent huge time gaps, or hiatuses (Figure 16). In fact, he was convinced of their existence even before he found the first example at the renowned Siccar Point in Scotland. John Playfair, an ardent supporter, was quick to

Figure 16. The renowned angular unconformity at Vallis Vale, in Somerset, which helped to illuminate the immense scale of the earth's rock cycle to Victorian England. The lower rocks are gray coralline limestone from the Carboniferous Period. Following deep burial, the limestone was folded by enormous horizontal pressures when the Rheic Ocean closed during the Variscan orogeny. The now tilted Carboniferous layers were then thrust upward into Himalayan-scale mountain chains that dominated western Europe at that time. After those mountains had eroded back down to sea level 170 million years later, further inundation by the sea and subsidence of the land during the Jurassic Period led eventually to deposition of the horizontal, yellowish limestone layers above. (Philip Bishop/Alamy Stock Photo)

recognize their significance, remarking that "the mind grows dizzy staring into the abyss of time." What Hutton had succeeded in conveying to his friend was that inconceivably huge spans of time are required to form an angular unconformity. First, soft sediments need to be buried deeply enough by later deposits for them not only to turn into rock but to be able to deform plastically under immense subterranean pressure. The now folded rocks must then be raised into mountain chains by the same crushing forces, before being slowly eroded down to sea level where they can finally be carved by the ebbing of tides and the crashing of waves. Angular unconformities represent, therefore, the closing of an immense rock cycle that, following further subsidence, can begin anew with another cycle of deposition. The fact that the Jurassic sediments at Tedbury Camp have likewise turned to stone indicates that at least one further cycle of burial, uplift, and erosion has taken place there since then, possibly related to the collisions that formed the Alps. Hutton recognized in such phenomena a restless cyclicity within the seemingly infinite pedigree of earth history.[3]

The angular unconformity at Tedbury was caused by colossal tectonic events to the south, namely the Hercynian orogeny, or mountain-building event. This was a time when gigantic Himalayan-scale mountain chains arose right across western Europe after the continent of Euramerica had collided with Gondwana to form the supercontinent Pangaea. Supercontinents have formed episodically since Archean times whenever most of the earth's continental crust has managed through a series of collisions of smaller fragments to amalgamate into one large block, or craton. Clearly, there are no mountains in the west of England anymore, but this does not mean that tectonic events from long ago have no consequences today, because unconformities juxtapose two very different ages and types of rock, and in this case as elsewhere in western Europe this juxtaposition would have important consequences for the Industrial Revolution.

Getting one's head around angular unconformities is still a major challenge for beginning geology students, as it was for me when I was an undergraduate of the Royal School of Mines in London. For many people, it will be the first time that they have thought about how our planet changes through the four dimensions of space and time. It may be their first attempt to grasp what it really means that mud turns to rock over time, rock turns to putty with pressure, and mountains dwindle to nothing by the incessant gnawing of the elements. The first question I ask students on arriving at Tedbury Camp has nothing to do with such things but instead is simply whether, by any chance, they noticed the names of the narrow country lanes our expert coach driver had to negotiate, sometimes with great difficulty, to get here. After turning off the main road, the first lane we take is called Iron Mill Lane, although it is not immediately evident how a tiny country road deserves such a big name. From Iron Mill Lane we then turn into Coal Ash Lane before arriving in the tiny hamlet of Great Elms. Even the name of the house we pass on our way to the quarry evokes a very different past: Steel Mills Cottage has an idyllic setting next to a stone bridge over a gently babbling brook. Nothing could be further removed from the cranking noises and sooty clouds of coal-fired iron mills, but make no mistake, this beauty spot was once a heartland of Britain's Industrial Revolution. Just one overgrown site in nearby Mells employed more than 250 men in its heyday forging agricultural equipment. But why here and not in, say, London, surely a much better connected metropolis? The ages of the rocks provide some clues.

The Carboniferous Period of the international geological time scale was so named because Europe's coal deposits are hosted in rocks of that age. The first forests appeared during that time, leaving behind their leaves, roots, and even trunks as fossilized organic remains. Plant fossils can be so abundant that in places they make up thick, mineable layers, and the Carboniferous rocks of Somerset are no exception. Near the spindly coral reefs of Tedbury Camp, stagnant coastal lagoons provided the perfect setting for the deposition of plant-debris-laden sediments that a combination of time, heat, and pressure turned into black gold. Twenty coal mines once operated directly west of here, remnants of a more widespread local industry that began in Roman times. Coal was, of course, a vital resource during the Industrial Revolution, but rocks are expensive to transport, so early forges and mills were located where coal could be found next to a source of running water and lime. At the start of the Industrial Revolution, before the railway network became established, lime for mortar, plaster, and concrete had to be made locally in lime kilns, where mainly Carboniferous limestone would be decomposed to calcium oxide at temperatures in excess of 900 degrees Celsius. The local iron works, and even great buildings like Glastonbury Abbey and Wells Cathedral, stand as monuments to the regional unconformity, having been built from blocks of Jurassic oolite cemented by lime produced in kilns fired by Carboniferous coal.

The Jurassic Period is also famous for the huge incumbents of Jurassic Park, some of whom must have sauntered hereabouts, as dinosaur footprints have been found in rocks of this age in nearby Dorset. When the sea finally encroached on the land, it left behind remnants of the wave-swept beach in the form of "borings" (tube-like holes where lithophagous, or rock-eating, worms and bivalves have etched into the beach rock just like they do today), oyster shells, and strange little calcite balls called ooids, which can be found today in the Bahamas rolling around like egg-shaped sand grains in balmy but turbulent waters. We met ooids already in a previous chapter because they formed in Scotland shortly before the onset of Snowball Earth. The Jurassic oolites of England were used to create some of the most famous buildings in the world, such as St. Paul's Cathedral, the UN Headquarters in New York, and the colleges of Oxford, Cambridge, and London. The fossil snails, oysters, and ammonites of the Jurassic coast can now be seen on polished walls and floors all over the world. The soft oolite at Tedbury Camp was stripped away in preparation for the blasting of the much harder Carboniferous limestone. However, for economic reasons the plan was not feasible, and the quarry workers left the horizontal contact between the two limestones intact. Anyone can now walk across this once coastal platform to retrace in their mind's eye the footsteps of a family of cetiosaurs, England's largest known sauropod dinosaur. An enormity of time, 170 million years, separates our footsteps from theirs. Curiously, the same amount of time separated their footsteps from the lithified coral reefs they were walking on. It turns out that 170 million years is about average for a rock cycle.

I can't say we ever achieve the same eureka moment with our students that Hutton clearly managed with Playfair. I remember well my own undergraduate field trips and know first-hand that it takes a lot of imagination on a cold, wet winter's day with failing sunlight to eke out the mind-blowingly profound from the seemingly banal. However, I hope that such trips do broaden the mind beyond the merely visible, and it is fortunately the case that field trips are usually among the more fondly remembered bits of geology degrees. What angular unconformities, such as at Tedbury Camp, show is that weathering and deposition lie at opposite ends of grand tectonic cycles, separated by, in this case, at least 170 million years. As will become a common theme throughout the rest of the book, the huge disconnect in time between storage and release can lead to tremendous imbalances in elemental budgets in the oceans and atmosphere. Crucially, this includes the earth's oxygen budget because deposition and weathering are the major sources and sinks of oxygen, respectively, on geological time scales. Oxygen is released when organic matter escapes the normal processes of oxidation and decay by becoming part of the rock record, for example as coal. Conversely, oxygen is consumed when fossils are eventually released from their rocky tombs and exposed to the atmosphere during weathering. Hutton's rock cycle can, by virtue of its immense scale, drive global compositional change, potentially creating permissive environments for preconditioned life forms to seize the day. Is it possible that an increase in the earth's oxygen budget allowed aerobic metabolisms to thrive, perhaps newly spawned in hyper-oxygenated cryoconite holes?

It takes an unimaginably long time for tectonic uplift to return huge volumes of deposited material back to the surface and be exposed to the elements once again. Ocean sediments take a similarly long journey from basin centers to rims, where, following subduction, some portion may reemerge after the uplift and erosion of volcanic arcs, closing another tectonic cycle. These gargantuan cycles of deposition and weathering imply that potentially valuable nutrients, like phosphorus, remain chemically unreactive for literally ages before they fleetingly reenter the active biogeochemical cycle. Take carbon as another example. The carbon content of coal can be greater than 90%, and even the small Somerset mines produced more than a million tons of coal every year in their heyday. Multiply that by all the fossil organic matter in the world and it becomes obvious that an enormous amount of carbon is stored in rocks. Burning coal releases almost four times its weight in carbon dioxide, consuming in the process the exact same number of oxygen molecules that had once been released by photosynthesis. Once locked up in solid form, fossil carbon can only be oxidized back to carbon dioxide, either slowly during oxidative weathering or more rapidly by burning, an activity that humans have greatly accelerated since the Industrial Revolution.

This long-term organic carbon cycle complements the inorganic carbon cycle we looked at earlier, though we did not explore in detail its consequences for the oxygen we breathe.

As long as the carbon remains locked up in rocks, the oxygen that was released during photosynthesis will simply stay in the atmosphere. Each year, photosynthesis just about matches organic decay, but some plant material will inevitably escape the rot, keeping oxygen in the surface environment. In the case of the Carboniferous coal deposits of Somerset, the oxygen released by their formation is only now, 340 million years later, being taken back, through a combination of natural weathering and coal burning. Millions of years of this carbon leak from the "short-term carbon cycle" to the long-term carbon cycle explain how our atmosphere has remained continuously habitable for animals from their very first breaths to ours today.

Most of the oxygen that organic burial gives, oxidative weathering will—eventually—take away, thus completing the long-term organic carbon cycle. However, this cannot be all of the earth's oxygen story, because if everything was always in such perfect balance then nothing would ever change on geological time scales. Moreover, there really needs to be an overall oxygen production surplus because a small portion is used up when ancient landscapes turn red with rusted iron, or during the oxidation of volcanic gases, like hydrogen sulfide, which is continuously belching out from volcanoes.

It turns out that the sulfur cycle has a co-starring role in regulating oxygen. As with carbon, sulfur has both reduced and oxidized forms that can be either buried or weathered. Moreover, reduced forms of sulfur are more easily oxidized during weathering than organic carbon. A lot of organic-rich sedimentary rocks, especially marine sedimentary rocks, contain reduced sulfur in the form of organically bound sulfur or mineral sulfide, which typically consists of golden cubes or smaller raspberry-shaped framboids of pyrite (iron sulfide). Pyrite typically forms under anoxic conditions when sulfate-reducing microbes catalyze the decay of organic matter, releasing hydrogen sulfide that can react with reduced (ferrous) iron to produce pyrite (FeS_2). Following burial, crystals of pyrite will be protected from reoxidation until tectonic uplift and erosion expose them to weathering. Oxidation of pyrite is quick and easy beneath today's oxygenated atmosphere. Indeed, the inadvertent presence of any sulfide mineral in concrete will weaken buildings beyond repair, as any structural engineer knows. In the same sense that the net effect of organic burial is to "release" oxygen, pyrite burial also allows free oxygen to remain in the environment for one rock cycle. For every molecule of sulfate reduced, almost two molecules of oxygen are freed up, whereas the converse is true when pyrite undergoes oxidative weathering. Atom for atom, the sulfur cycle is almost twice as important for balancing our oxygen budget as the carbon cycle (Figure 17).[4]

Convention dictates that any surplus in one element's oxygen balance (burial versus weathering) must be compensated for by a deficit in the other element's balance. For example, if more organic carbon gets buried than weathered, thereby causing a net increase in oxygen, then that might cause a fall in pyrite burial, which would help to rebalance the earth's oxygen budget. This balancing act can perhaps best be seen following the emergence of the first forests during the Devonian to Permian Periods. Terrestrial environments where coal deposits form are often rich in carbon, but are generally poor in sulfur, thus providing a perfect mechanism whereby higher (terrestrial) organic burial could be compensated for by lower (marine) pyrite burial. This coupling is evidenced in seawater sulfate concentration, which rose markedly during this interval, resulting eventually in gypsum (calcium sulfate) precipitation throughout the Permian Zechstein Sea of northern Europe. In redox terms, surplus carbon reduction is generally compensated for by surplus sulfur oxidation, allowing the earth's oxy-

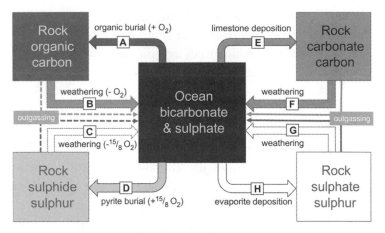

Figure 17. Box model of the earth's surface oxygen budget,
showing the flux (A through H) of carbon and sulfur through
the exogenic (surface) earth system. On long time scales E = F,
but there can be imbalance in the sulfur cycle whereby G ≠ H
because flux G leads instead to D, releasing oxygen that can be
used to increase flux B, causing negative carbon isotope excursions
(as explained in chapter 10). The net oxygen gain to the surface
environment = $(A + {}^{15}/_8 D) - (B + {}^{15}/_8 C)$.

gen budget to remain steady (and habitable for complex life).
A rough negative relationship over time between seawater car-
bon isotope ($^{13}C/^{12}C$) and sulfur isotope ($^{34}S/^{32}S$) ratios has been
attributed to this delicate balancing act. Although geochemists
don't imagine that this negative feedback works on very short
time scales, it is conceivable that it may not always work even
on longer time scales. We learned earlier that Snowball Earth
events were followed by extraordinary changes to chemical
weathering, which would have flooded the world's oceans with
nutrients, fueling organic production and releasing oxygen.
Driven by high temperatures and high carbon dioxide levels,
rather than tectonic uplift and erosion, any increase in organic
carbon or pyrite burial at those times would not have been com-

pensated for by an increase in the oxidative weathering sink. It would indeed be surprising were those catastrophic weathering events not to have had major consequences for oxygen. But how do we go about proving this?

Three years after my first trip to the Gobi-Altai ranges of western Mongolia, I was back again for more, but this time on my own. After defending my thesis in Zurich, which in Switzerland is quite a public and—for me—chastening experience, I flew straightaway to Beijing, where the International Geological Congress was about to begin. I was due to present a paper boldly entitled "Global Environmental Change Following the Late Precambrian Ice Ages: A Rare Earth Element Approach." In the talk, I outlined the gist of an idea that I had circulated on the web earlier that year in the fledgling "Internet Journal of Science." It was 1996, the heady days of the World Wide Web, when new websites hatched, flew, and crashed before anyone really noticed. You can still find the abstract for the talk I gave on microfiche (younger readers may have to look that word up), but the accompanying online paper has disappeared without a trace. Following the conference, I bought a ticket for the Trans-Siberian railway in the hope of securing a visa after arrival. Western Mongolia was in lockdown, a familiar term nowadays, due to a nasty combination of black plague and cholera. The plague is endemic to central Asia due to the unfortunate habit among nomadic hunters of bringing back freshly slain marmots to their family yurts (or *gers* in Mongolian) for skinning. By the time they arrive home, the fleas are desperate to escape their now cold host, thus spreading the disease. I remember queuing up to cajole the Ulaanbaatar officials into granting me extension after extension. I was still not permitted, however, to travel outside the capital but managed to use my expired student ID and, failing that, the occasional bottle of vodka to gain free passage.

In Beijing, I had reported that the Sturtian ice age in Mongolia was followed by progressive enrichment of the heavier isotopes of both carbon (^{13}C) and strontium (^{87}Sr). Several years earlier, similar trends had been shown to follow the later Marinoan event by (Alan) Jay Kaufman, in Namibia. A key member of the renowned isotope laboratory at Harvard University, and a key collaborator of Andy Knoll's, Jay published the geochemical evidence for the resurrected Snowball Earth hypothesis just two years later in Paul Hoffman's now classic paper. In chapter 5 I described why rising seawater ^{87}Sr/^{86}Sr ratios are consistent with enhanced chemical weathering in the post-glacial greenhouse. Although increases in the ^{13}C/^{12}C ratio (or δ^{13}C) are usually attributed to high rates of carbon burial, the amount of oxygen released by organic burial cannot be quantified, Jay pointed out, if we do not know how much carbon was passing through the earth system. By ingeniously combining these two systems, he was able not only to estimate relative changes to carbon burial (and resultant oxygen release) but also, by constraining erosion rates, to quantify absolute changes to carbon burial. Taken at face value, our two sets of results were consistent with the idea that the surface environment became more oxygenated after each Cryogenian ice age. These findings were important because we cannot measure past oxygen levels directly, handicapping greatly any attempts to decipher how oxygen might have affected life, or conversely how life might have affected oxygen. Nevertheless, the isotopic evidence for oxygenation was still more indicative than conclusive. We needed to find an independent means to demonstrate the presence and amount of oxygen in the post-glacial environment; hence my interest in the rare earth elements, or REE.[5]

The rare earth elements are a group of fourteen naturally occurring heavy transition metals and are such a tight-knit band of chemical siblings that they are virtually inseparable

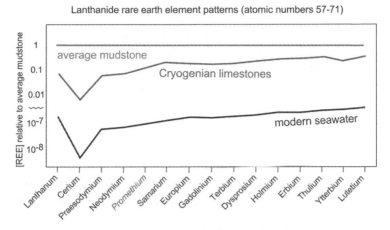

Figure 18. Representative rare earth element (REE) concentrations in ~660-million-year-old post-glacial Cryogenian limestones from western Mongolia compared with modern seawater. Cerium is the only rare earth element that oxidizes to an insoluble state, causing a relative deficiency of cerium beneath an oxygenated atmosphere and ocean. Geochemists use changes in these "negative cerium anomalies" to trace oxygen through time.

(Figure 18). Although their names are hardly household words, the REE have been in the news a lot lately due to their importance for the tech industry. Rather than being rare, they are fairly commonplace in the earth's crust but seldom found in the abundances needed to justify starting a mine. As a result, the REE have become a geopolitical football, with China enjoying a near monopoly on the scarcer and more valuable heavy REE. As they generally exist as large 3+ ions in nature, the REE all react in the same predictable ways, or at least they did until after the Great Oxidation Event, when cerium, the most abundant REE, could be oxidized into a much smaller 4+ ion, thus falling out of solution. This process is akin to the rusting of iron, which occurs when iron oxidizes to insoluble Fe^{3+} ions, although Ce oxidation occurs at a higher oxidation state. Modern oceans reveal cerium oxidation as a "negative Ce anomaly," meaning that cerium is anomalously scarce compared with its REE siblings. Precisely when the oceans became sufficiently oxidized to show a modern-sized Ce anomaly is a moot point, but by the mid-1990s no major anomaly had yet been detected in any Precambrian sedimentary rocks. My data showed a clear trend to substantial Ce depletion during the aftermath of Sturtian glaciation, when strontium isotopes implied greater chemical weathering rates and carbon isotopes indicated increased organic burial. In my talk, I speculated that this covariation signaled a globally important rise in atmospheric oxygen after deglaciation. The Mongolian samples are still among the most cerium-depleted of any rocks older than the Devonian Period.[6]

Stepping off the Trans-Siberian at Ulaanbaatar, my only plan was to get back to Altai to check those findings. Seizing the first opportunity that came along, I hired a minibus, driver, and translator to take me there and back. On the first trip, we miraculously gained two students from Germany, an Israeli who had been traveling over the steppe on horseback, a quarreling couple from Liechtenstein, and a young boy whose nomadic family was sending him to the capital to go to school. During the second trip, we detoured to the Golden Cliffs of the Gobi Desert, where we could see the skulls of *Protoceratops* dinosaurs, polished white by sandstorms, eerily looking up at us from beneath the sand dunes. Logistics and lockdowns meant that it was not until the third trip that I could finally retrieve some more samples to complement the earlier set, and months later, back in the laboratory, we were able to replicate the results of three years earlier. Questions remained, however, over whether this oxygenation was a regional or a global phenomenon. The correlation between tracers of global processes, like strontium isotopes, with local phenomena like cerium depletion implied a common origin, but more evidence was needed to support the Neoproterozoic Oxygenation Event, or NOE, hypothesis. The hunt was on for other so-called redox-sensitive, or multivalent, metals that would be sensitive at different oxidation potentials.[7]

Many elements can exist in several different oxidation, or valence, states, and in the ensuing years numerous attempts have been made to find the geochemical smoking gun for the NOE. The best tracers appear to be transition metals, such as chromium, vanadium, uranium, and molybdenum, that can exist in two or more oxidation states. All four elements are abundant in today's oxygenated oceans because when fully oxidized they form strong bonds with oxygen, in the form of highly soluble oxyanion complexes. However, they tend to become less soluble in their reduced forms, and so are sequestered into the sediment under anoxic conditions. Studies of the Marinoan aftermath confirm extremely high concentrations of these transition metals in China, indicating local anoxia. Crucially, their high concentrations imply that the global ocean was, by contrast, widely oxygenated. No such high enrichments have been reported in older rocks because, presumably, a much greater expanse of the seafloor was anoxic before the Ediacaran Period. Elements like molybdenum are exceptionally sensitive to the presence of aqueous sulfide, so their enrichment indicates more specifically that local waters were not only anoxic but sulfidic, and that the global ocean was widely oxygenated and exceptionally rich in oxyanions like sulfate.

Many subsequent studies have provided evidence for progressive but episodic oxygenation throughout the Ediacaran Period, resulting in near modern-like conditions during Cambrian times.[8] When interpreting geochemical data in terms of environmental "redox conditions," we need to remember that each elemental or isotopic tracer addresses a different question.

First, each element has a different range of conditions over which they are likely to be sensitive, so some tracers will respond to change when others might not. Moreover, some tracers reflect global conditions, where others are only regional or local, due to differences in their ocean residence times. Some tracers follow oxygen directly, while others trace it indirectly through oxyanions like sulfate or molybdate. Some tracers respond to atmospheric oxygen, whereas others trace the extent of ocean or seafloor anoxia, and so on. It is conceivable, for instance, and perhaps even to be expected, that different tracers will flatly contradict one another, resulting in considerable confusion, or nuance, depending on your point of view. Changes in atmospheric oxygen, for instance, may take an entirely opposite trajectory to ocean oxygen. This is because free oxygen can only build up when its sources overwhelm its sinks, which are different on land than in the sea. The ultimate source of most oxygen on earth is photosynthesis. Whenever an atom of organic carbon is buried, a molecule of oxygen, the waste product of photosynthesis, gets left behind. The released oxygen can either build up in the environment or, more likely, react with an oxygen sink. There are many such oxygen removers because oxygen is just so reactive. Free oxygen reacts every day with volcanic gases, methane, ammonia, pyrite, decaying organic matter, and all sorts of transition metals that, like iron, manganese, or cerium, can be "rusted" into a higher oxidation state.

Oxygen production today can easily enough overwhelm terrestrial oxygen sinks and keep our atmosphere and oceans well oxygenated. Even today, however, it is possible for parts of the oceans to turn anoxic even though the atmosphere is highly oxygenated, especially where nutrients are abundant. This is because oxygen sinks, such as ferrous iron, sulfide, and organics, can accumulate in the ocean, sapping oxygen in areas where organic production is high. Bodies of seawater with free sulfide are called euxinic, after the Roman name for the Black Sea, whereas oceans rich in reduced iron are referred to as ferruginous. We don't, but really should, have an equivalent word for water bodies containing lots of decaying organic carbon. As we saw in chapter 6, the Precambrian ocean may have had all three of these oxygen sinks (ferrous iron, sulfide, and organic carbon) in abundance, albeit with considerable variation through both time and space. To appreciate such nuance, we need to combine the above elemental and isotopic tracers with one of the most useful involving iron. Because iron has a very short ocean residence time under both oxic and sulfidic conditions, its behavior is extremely sensitive to local seafloor redox conditions.

The Great Oxidation Event (GOE), introduced in chapter 6, was the time when oxygen first arose in the atmosphere. The absence of detrital pyrite in any post-GOE sediments assures us that atmospheric oxygen crossed a threshold of about 0.1% of present levels around 2.4 billion years ago and has exceeded that threshold ever since. In other words, since the GOE, all pyrite exposed to weathering has rusted away to iron (III) and sulfate ions, which rivers then transport to the oceans. However, this is not the end of the marine iron cycle. Iron (III) is insoluble, so although it enters the ocean as particulate matter, it can still be reduced back again to more soluble iron (II). Today this occurs almost exclusively within the sediment, leaving the seafloor fully oxygenated. However, anoxic oceans can accumulate dissolved iron and become "ferruginous." Most of the world's iron ore derives from enormous "banded iron formations" that precipitated out from such ferruginous oceans between about three and two billion years ago.

Fe (II) is therefore extremely sensitive to ambient redox conditions. In the presence of even a tiny amount of free oxygen, it is liable to lose a further electron to become Fe (III) and fall out of solution. If this happens on a big scale, then iron concentrations will be very low, as in today's oceans. In the presence of sulfide, Fe (II) is liable to titrate out of solution in one of a number of sulfide minerals, including pyrite. When sulfate and reactive organic matter are freely available, the dominant control over pyrite precipitation will simply be the availability of iron. Therefore, under euxinic conditions, pyrite will dominate the pool of reactive iron in a sediment. Where conditions are anoxic, but not sulfidic, iron is free to build up, leaving the sedimentary iron inventory dominated by other mineral types, such as oxides or carbonates. This is the essential principle behind what is popularly known as "iron speciation," although it should more correctly be referred to as iron phase partitioning, a technique developed painstakingly for modern marine environments and then applied to the ancient world by a series of renowned geochemists from Bob Berner, Rob Raiswell, and Don Canfield to Tim Lyons. In more recent years, however, iron speciation has been the indisputable domain of Simon Poulton of Leeds University, who has been singular in his efforts to hone iron speciation techniques. Simon's extraordinary work has shown that ancient marine environments can, using empirically defined pools of extractable iron, be neatly divided into oxic, ferruginous, and sulfidic realms. Iron speciation identifies local to regional seafloor redox conditions, but more importantly shines a light on how early animals made their living (Figure 19).[9]

Toward the end of 2006, I was contacted by Guy Narbonne, a renowned Canadian paleontologist and leader of the memorable excursion to Newfoundland just a few years earlier. Guy wanted to share news of a study he was involved in that was about to appear in the journal *Science*. The paper concerned a very subtle change in the types of iron minerals before

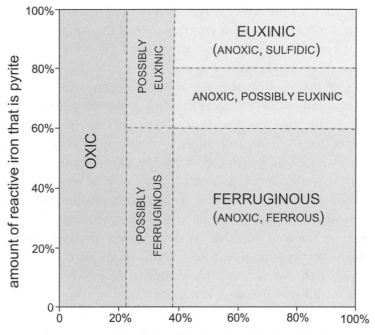

Figure 19. A classic "Iron speciation" diagram. Iron
partitioning between different mineral phases in a marine
sediment is determined by whether the overlying and
neighboring water column contains oxygen or not, and
if not, whether it contains sulfide.

and after the mid-Ediacaran Gaskiers glaciation of around 582
million years ago. It was one of the first papers to be published
using the more exhaustive iron extraction method that had
been developed by Poulton and Canfield. The new study made
the outstanding claim that the deep marine environment became oxygenated right at the moment when Ediacaran rangeomorphs began to populate the seafloor at Mistaken Point in
Newfoundland. Although not a truly global event, the Gaskiers

glaciation has now been recognized on several other continents, and it could be that this event represented the "smoking gun" of the earth's second great oxidation event, the NOE. The iron speciation data supported an earlier claim by Don Canfield in 1998 that the GOE had affected the atmosphere and surface ocean only, while the deeper realms remained anoxic until much later. Just how late was a surprise to everyone, and the paper had a suitably large impact, triggering an avalanche of independent studies in support of the twin hypotheses that the Ediacaran biota had aerobic lifestyles and so were likely to have been animals, and that they radiated throughout the deep marine environment at a time of ocean oxygenation.[10]

At the same time that forests of rangeomorphs inhabited the newly oxygenated Avalonian continental shelves, in China an altogether more hostile environment could be found. During the Ediacaran Period, the South China seaway was sandwiched between the Yangtze and Cathaysia microcontinents. Chinese rocks of this age are extremely rich in organic matter and pyrite for much of this time. Iron speciation indicates widespread euxinic conditions with occasional incursions of ferruginous or even oxic water. Yet even a fleeting presence of oxygen seems to have been enough to populate the South China seafloor, albeit very briefly, with multicellular algae and the enigmatic, eight-armed *Eoandromeda,* which has also been discovered in Australia. As with the immediately post-glacial enrichments I mentioned earlier, these euxinic conditions favored the removal of sulfide from seawater, and with it lots of other redox-sensitive trace metals, such as molybdenum, vanadium, and uranium. Our work in China has suggested very high concentrations of molybdenum in the Ediacaran and Cambrian black shales, supporting the evidence for euxinia, but, more importantly, suggesting that the global reservoir of oxyanions had risen in stages to near modern levels. Final confirmation of an oxygenated ocean during the Ediacaran Period has come from a range of metal isotopes, including, again, molybdenum, but also uranium, selenium, and chromium.[11]

At this point, the preceding description may well seem confusing. Redox tracers suggest oxic conditions in Newfoundland and anoxic conditions in China, but somehow I am arguing that both data sets are consistent with an oxygenating marine environment. This apparent contradiction goes back to the earlier point that oxygen derives from the production and storage of organic matter and/or pyrite, which requires anoxic conditions somewhere in the world. It seems that China was one of several highly anoxic, possibly restricted seaways, which helped the oxygenation of the deep oceans elsewhere, just as highly productive areas of upwelling, such as off the coast of Namibia, do today. What might seem at first to be a contradiction is actually a necessary juxtaposition of local anoxia, promoting global oxygenation. Clear as mud? I hope not, but, if so, then later sections will help to make this point much clearer.

It would be appealing to imagine that as soon as oceans became more generally oxygenated it was "Hello, modern world" without looking back. After all, muscular animals have been around ever since they evolved during the Ediacaran Period, and we all know that an energetic lifestyle needs plenty of oxygen. Unfortunately, it isn't really that simple. Whereas molybdenum isotope data tell us that the amount of seafloor anoxia had declined to modern levels already by the early Cambrian, iron speciation studies show us that anoxia could still be remarkably widespread at times, and wholly unlike the modern world we live in today. It seems that oxygenation was localized, transient, or both throughout the Ediacaran-Cambrian transition interval. Such nuanced narratives are fuel to the fire of the permissive environment skeptics, chief among whom is Nick Butterfield, mentioned earlier as the discoverer of the oldest red alga in the world.

Nick is an entertaining speaker who is sometimes expressly asked to speak at conferences, sometimes directly after me, to spoil the geochemists' party. The counterweight that skeptics like Butterfield provide is extremely important, as we can only be sure of our knowledge once we examine all the alternatives. It pays to keep an open mind. The oxygen debate evokes memories of Daly's Ploy versus Sollas's Gambit, mentioned in chapter 7, although whether it was environmental change or biological evolution that triggered the Cambrian explosion now centers on oxygen rather than biomineralization. It is important to note that Butterfield's critique of the "oxygenation led to animal evolution/radiation" hypothesis does not rely on geochemical evidence, as he seems equally happy with either no change in oxygen levels through this interval or with significant oxygenation, but of purely biological origin. This latter position is shared by many biologists who seem to consider evolutionary innovations as the main drivers of environmental change. However, I do not think that such a claim is supported by the evidence.

The argument goes like this. It took an awfully long time for eukaryotes to evolve the requisite genetic toolkits for developing both multicellularity and heterotrophic metabolisms. Once everything was in place, an arms race ensued that allowed animals to diversify into a plethora of body plans, which became our modern animal phyla. Animals likely oxygenated marine ecosystems by packaging organic detritus into heavier ballast in the form of excrement and by cleaning up organic detritus by filter and suspension feeding or direct ingestion. The clearer and more oxygenated oceans finally allowed eukaryotic algae to compete with cyanobacteria, thus transforming the oceans from the stagnant and turbid microbial stew they had been previously. There is a lot to be said for this view of the world, and something along these lines very likely happened. However, it is equally true that one cannot clearly separate biological from geological change, as the one almost certainly affects the other. Ever since aerobes evolved, they would have been susceptible to times of oxygenation and deoxygenation, so it makes little sense to deal with biological evolution in a vacuum. More pointedly, without permissive geological drivers, biological innovations would be useless because the oxidizing capacity of the surface environment is at the mercy of Hutton's gargantuan rock cycle of deposition and weathering.[12]

Considering this, it seems unlikely that the rangeomorphs, or any other types of primitive animals like sponges, can take the credit for oxygenating the oceans during the Ediacaran Period. This is because their aerobic metabolism would have sapped oxygen from the environment. Any success they might have had would therefore have been self-defeating in a low-oxygen world, unless some overarching driver was producing surplus oxygen. It seems counterintuitive to me to believe that the activities of newly evolved organisms that consume a given resource, like oxygen, would by their actions increase the availability of that resource. In one regard, though, animals would have increased the availability of oxygen. The packaging of organic matter by guts into fecal parcels and the bodies of sessile filter-feeding or osmotrophic organisms would have relocated the sites of oxygen consumption to deeper realms and to the seafloor, helping shallower waters to clean up their act. In other words, redox boundaries in the Precambrian environment would have been more diffuse, with oxygen consumption spread out over a much thicker water column. However, sharpening this redox boundary would not directly change the size of the marine oxygen budget, just redistribute it. The size of the oxygen flux into the surface environment is determined primarily by the production of organic matter, and resultant organic carbon and pyrite burial, which in turn is determined by the availability of nutrients. We will turn to nutrients in the next chapter, but suffice to say that current thinking suggests animals may even have diminished the world's oxygen production by more effectively removing the major limiting nutrient, which is phosphorus. They achieved this by irrigating the seafloor and consuming buried organic matter by their burrowing activities as well as by directly storing phosphorus in their skeletons in the form of apatite, just as we humans still do today.[13]

In some sense, this is an artificial debate because geochemical evidence for at least episodic oxygenation is so strong, and carbon isotopes even allow us to quantify the extent of oxygenation events, as we shall see later. However, when we discovered that along with evidence for oxygenation there was ample evidence in the opposite direction, this was the final nail in the coffin for a biologically driven oxygenation. Rather than heralding the dawn of a new oxygenated surface environment, the rise of early animals apparently was accompanied by precisely the reverse trend during the transition to the Cambrian Period. Instead of being a stepwise shift in baseline oxygen levels, the transition to modernity seems to have been a very noisy affair indeed, with extremes in both directions. It had been known since the mid-1980s that this entire interval was marked by extremely high amplitude carbon isotope excursions, but it was not until analysis of other isotope systems, like uranium, became possible that we could finally show how these carbon cycle fluctuations matched pulses of ocean oxygenation. This exciting leap in understanding has come about only in the last few years.

Rachel Wood, a professor of geology in Edinburgh, has spent a large part of her career looking at carbonate rocks. On my first trip to Mongolia, Rachel was already well known for her work on reefs and spent much of her time on that trip examining the vase-shaped skeletons of calcareous archaeocyath sponges, once thought to be the first animal reef-maker. More recently, Rachel and her student Amelia Penny have described the oldest Ediacaran-age, reef-building animals from Namibia, in the form of cloudinids, which are calcareous "cone-in-cone" tubes of uncertain taxonomic affinity (Figure 20). These days, with infectious enthusiasm, Rachel leads her team almost every year into the Namibian desert, one of the best places in the world to view the Ediacaran-Cambrian transition to the modern biosphere. Rachel's group has discovered not only the world's earliest animal-based reefs, but also some of the earliest evidence for true locomotion and brainy behavior in the form of fossilized trails and burrows. The Namibian fossils derive from a crucial juncture in earth history between the rangeomorphs of Avalonia (southern British Isles and Newfoundland) and the Cambrian radiation of modern animal phyla. If evidence for the relationship between oxygen and evolution can be found anywhere on earth, it ought to be found here.[14]

Contrary to all expectations, iron speciation studies indicated that the Namibian shelves, although teeming with animal life in places, were widely anoxic, even though they were fairly shallow. The Namibian reefs, it seems, were in fact oases of oxygen surrounded by lethal anoxia. Something had clearly changed since the deep ocean oxygenation just ten million years earlier. As with China, we might try to explain this as due to high levels of nutrients and organic production in a more generally oxygenated ocean. This possibility, however, was challenged by Rosalie Tostevin, now of Cape Town University, who applied a range of different geochemical tracers to tease apart local and regional from global redox conditions. Rosalie, at the time a PhD student working with Rachel and me, first set about

using the REE, which quickly confirmed that only the upper-most Namibian seas were fully oxygenated. Everything deeper was either suboxic or entirely anoxic. With the help of Clau-dine Stirling, an isotope geochemist in New Zealand, Rosalie was able to add uranium isotopes to the mix with astounding results. Although mid-Ediacaran oceans were clearly well ox-ygenated by all accounts, late Ediacaran oceans were just the reverse, with equally extreme uranium isotope ratios but in precisely the opposite direction. Although a single study might be dismissed, this shift has now been detected in China and Mexico as well. The Ediacaran-Cambrian radiation of modern animals seems to have been accompanied by unprecedented swings in global oxygen availability.[15]

At the same time that Rosalie was looking at the latest Ediacaran ocean, another UCL student, Tianchen He, was busy extracting similar information about the Cambrian ocean, using a combination of carbon and sulfur isotopes. It has long been known that this interval is exceptional for its wild swings in $\delta^{13}C$, but Tianchen discovered that each carbon isotope cycle was matched by an equally exceptional swing in sulfur isotopes,

Figure 20. (*opposite*) Two characteristic features of the Cambrian explosion: biomineralization (top) and bioturbation (bottom). The world's earliest animal-based reefs formed around 555 million years ago from nested tubes of calcite, like these exquisitely silicified examples from Spain. Such *Cloudina* shells, up to 2 millimeters in diameter and 20 millimeters in length, are thought to have been made by polychaete worms. The base of the Cambrian Period at 539 million years is marked by the appearance of the first complex burrowing behavior. These treptichnid burrows from Spain are thought to have been made by priapulid worms and are typical of the trace fossils that define the base of the Cambrian Period worldwide. (Photos by Ying Zhou, 2019) or $\delta^{34}S$. This made it unlikely that the oxygen release from organic burial was balanced by complementary changes to pyrite burial, meaning that each cycle corresponded to oxygenation and deoxygenation on a global scale. Before long, uranium isotope evidence again confirmed this instability, implying that the habitable space for these newly evolved aerobes was always changing. Quite possibly, such environmental challenges drove the exceptionally high rates of biological diversification that characterize the Cambrian explosion, with each episode of opportunistic radiation followed by habitat shrinkage and even extinction. Such extreme volatility could go a long way toward explaining Darwin's Dilemma concerning the unseemly abruptness of the Cambrian radiations.[16]

It is tempting to think that all these ups and downs are just noise and that there was no significant Neoproterozoic Oxygenation Event, but I think that would be wrong. Despite considerable volatility, the Ediacaran and Cambrian Periods saw huge increases in oxidizing capacity that culminated in a near modern extent of oxygenated seafloor by 520 million years ago. Despite fluctuations, the baseline trend toward a more oxygenated marine environment is clear. The ecological changes outlined by Nick Butterfield also ensured that vertical redox gradients of Phanerozoic oceans were fundamentally different from those of before. As we shall see later, the carbon isotope excursions alone imply enormous transfers of oxidizing capacity into the ocean-atmosphere system during the Ediacaran Period, and periodically throughout the early Cambrian, with biological events piggybacking on environmental changes at each rise. Such huge transfers of oxygen, or energy, imply equally huge transfers of material that can only relate to tectonic events of uplift and erosion, such as occurred to form the Tedbury Camp angular unconformity. The rock cycle of deposition and weathering has to hold the key here, but to dig deeper into this now would be getting ahead of ourselves. Although the picture is becoming ever more nuanced, there is plenty of geochemical evidence to show that environmental changes permitted the radiations (and extinctions) of aerobic organisms, preconditioned for a more oxygenated world, throughout Cryogenian to Cambrian times.

Whichever way you spin it, the production of oxygen through either organic carbon or pyrite burial is catalyzed by life. And life needs to be fertilized. Before we ask how the surplus oxygen got there—we will come to that in due course—we first need to ask where on earth the nutrients came from to fuel organic production and pyrite burial. As the sole nutrient that can come only from the weathering of solid rock, phosphorus is often referred to as the ultimate limiting nutrient. Indeed, phosphorus undergoes more cycling in the surface environment than any other element, even more than carbon or nitrogen, before being permanently removed into rocks. It may be no coincidence that so many of the world's phosphate resources, on which we rely for the fertilizer needed to feed the world's billions, are of Ediacaran-Cambrian age. Amazingly, these resources also contain some of the very best Ediacaran fossils, exquisitely preserved in mineral phosphate.[17]

9

Limiting Nutrients

The earliest putative animal fossils are petrified embryos, which, bizarrely, form one of the largest deposits of rock fertilizer in the world. Why so much phosphate formed after Snowball Earth is a long-standing mystery, likely related to weathering, productivity, and oxygenation. The ice ages took place on a long-lived, low-lying supercontinent that was slowly breaking apart, but their aftermath saw monumental collisions between rifted crustal blocks. The erosion of huge mountain chains accelerated biogeochemical cycles in waves of boom and bust for eukaryotic life forms throughout the Ediacaran and early Cambrian Periods.

A t least three-quarters of the oxygen flux to the atmosphere and oceans today is generated by organic carbon burial. The remaining oxygen derives from the burial of sulfide in the form of pyrite, following bacterial sulfate reduction. The relative importance of carbon versus sulfide in the global oxygen cycle may well have been reversed in the anoxic oceans of the past, as we shall see in chapter 10, but it is important to note that both of these oxygen sources are dependent on organic productivity, which is itself dependent on the availability of fertilizing nutrients. Phosphate is often said to be the ultimate limiting nutrient because it is so scarce in the natural environment, deriving solely from chemical weathering. Without phosphate there can be no life, let alone oxygen, and any scarcity will severely limit productivity, and therefore oxygen. Before we get to the implications of nutrient limitation for the earth's oxygen budget, allow me a short digression of a very human kind on the subject of phosphate.

In 2007, a large-scale riot took place in West Bengal, India, the first of many protests against the soaring cost of living, and especially of the three "f's": food, fuel, and fertilizer. Although widely regarded at the time as a protest against corrupt officials, the Bengal "food" riots marked the start of more widespread unrest. In April 2008, thousands of workers in Bangladesh went on strike over food prices, despite the country having been declared self-sufficient in food just a few years earlier. Rioting over the rising cost of such staples as rice spread farther afield, to Africa and then to Latin America, as people around the world increasingly struggled to feed their families. Government interventions caused prices to fall by the middle of 2008, but the problem did not go away. A second price hike four years later helped to fuel the Arab Spring uprisings, which led to civil chaos and the overthrow of four Middle Eastern governments.

In an effort to protect domestic markets, countries introduced protectionist policies as the food crisis wore on. However, as protectionism leads to fewer exports, such actions cause global commodity prices to rise even higher. For countries like Bangladesh that are strongly dependent on imported fuel and fertilizer to produce and transport food, global price hikes lie beyond the control of the government. Many of the African countries that witnessed food riots, such as Burkina Faso, Senegal, Egypt, and Morocco, which have major rock phosphate reserves for manufacturing their own fertilizer, still needed to import fuel, food, or both. In an increasingly populous, developed, and globalized world, few nations attain genuine food security and very few indeed can claim self-sufficiency in fertilizer.

Many commodity prices rose between 2007 and 2008, but none came close to matching the 450% rise in the cost of phosphate rock, or phosphorite. As a consequence, the price of phosphorus-based fertilizer skyrocketed. A recent analysis of the crisis concludes that the sharpest spike in the cost of phosphorite was caused by a drastic decline in fertilizer production in India. While other countries protected their internal markets by restricting exports, India was forced to double its importing of phosphorus fertilizer, further exacerbating the rise in global prices. The additional import demand for diammonium phosphate (DAP) has been estimated at about 20% of the annual global supply. India, like Bangladesh, simply does not have enough phosphorite to enjoy genuine food security.[1]

China also played a key role in the crisis. China uses a lot of fertilizer, but unlike India it has huge natural reserves, all formed during the Ediacaran or Cambrian Periods. At the time of writing, China mines more phosphorite than all other countries combined. Concerned by rising prices, China acted to prop up its domestic market by introducing an export tax of 100% on fertilizers. Other nations, such as Burkina Faso, behaved similarly to protect their domestic markets by subsidizing fertilizer costs and reducing exports. The combination of encouraging more fertilizer use while simultaneously limiting supply resulted in a classic positive feedback, driving global market prices ever higher and exposing food insecurity all over the world. Indigenous phosphorite supplies in India meet only 5–10% of national demand, making that country extremely vulnerable to price hikes. Other countries, including Bangladesh, have burgeoning populations but no mineable phosphate resources whatsoever. European agriculture is also dependent on imported phosphate. Phosphorus is essential for all life—so how can it be so scarce? What makes China lucky, but Bangladesh and so many other countries unlucky, in the fertilizer stakes? The answer is geological.

Although the nineteenth-century agricultural chemist Justus von Liebig usually gets credit for spawning today's mammoth fertilizer industry, his famous Law of the Minimum was actually a retelling of Carl Sprengel's earlier Theorem of the Minimum, which maintains that plant growth will be limited by the availability of the scarcest nutrient. Erasmus Darwin, grandfather of Charles, is often credited with the idea that bone meal, which is made of the mineral hydroxyapatite, a form of calcium phosphate, could be used to fertilize crops. Von Liebig maintained that micronutrients could also be important fertilizers but believed, wrongly, that plants could simply take nitrogen from the air, which meant that nitrogen could never be limiting. In fact, legumes are the only common plants that can do this, and only because of their symbiosis with nitrogen-fixing bacteria. It is the microbes that do the actual work.[2]

A result of trial and error, modern fertilizers include three major ingredients: potash, nitrate, and phosphate. Potash is a mixture of soluble potassium salts that precipitated long ago from evaporating seawater. Despite being abundant all over the world, its ubiquity makes it rarely profitable to mine, leading to near total monopoly of the global market by just three countries: Canada, Russia, and Belarus. Despite this, many countries in the world could at least in principle become self-sufficient with respect to potash. Nitrate tells a similar story, although it is not found in rocks at all but needs to be extracted from the air using the Haber-Bosch process, which mimics at great expense what microbes do for free. Although obtaining nitrate is extremely energy consuming, our atmosphere provides an almost inexhaustible supply of nitrogen. By contrast, phosphate can only be taken out of the ground and is much scarcer than potash. More than 90% of global phosphate production is dominated by just a handful of countries. Further, more than 80% of reported phosphate reserves, and well over half of all phosphorite mined each year, comes from just two countries, China and Morocco, and only from, respectively, Ediacaran-Cambrian and Cretaceous-Cenozoic rocks. But why are mineable deposits so sporadically distributed in both space and time? The answer lies at the interface of biochemistry and oceanography.

Phosphorus is required for all life on earth. Along with H, C, O, and N, P is an essential part of nucleotides, the building blocks of DNA and RNA. In the form of phospholipids, phosphorus occurs in all cell membranes, while as adenosine triphosphate, or ATP, phosphate hydrolysis unleashes the chemical energy needed to drive metabolic processes. For something so vital, however, phosphate is remarkably scarce in nature, so on geological time scales it has to be Sprengel's limiting nutrient. Indeed, it is so valuable to life that as soon as phosphorus becomes available, life consumes it as swiftly as possible, after which it either gets recycled sooner, through predation or decay, or later, once sediments have undergone burial, uplift, and weathering. Today, the phosphate ions released from weathering travel by river to the oceans, where, following organic production and decay, they may be released again at depth. In places where cold, deep waters rise, such as off the western coasts of continents, upwelling brings life-giving phosphate back to the surface, creating a boon for marine life.

Although all phosphate must enter the ocean via rivers, the riverine P flux is a thousand times smaller than the recycled P flux. Phosphate recycling, largely through ocean upwelling, creates the most productive marine realms on earth, where phosphate rains down onto the continental shelves in the form of bones, excrement, and tiny, nondescript particles of organic matter (POM), where it can accumulate in enormous abundance. When phosphorus is released during organic decay beneath the seafloor, some of it will end up locked in the sediment pile in a stable form of the calcium phosphate mineral apatite, called francolite, which is the major mineral constituent of phosphorite. Upwelling zones on the edges of the opening Atlantic and closing Tethys Oceans supplied much of the phosphate reserves that we now use to help feed the world's population. However, because the age distribution of phosphate reserves is patchy, it seems that such conditions were rare through earth history. By geological standards, the phosphorite deposits of the Mediterranean and North African region are not very old, but, surprisingly, there are only a few older phosphorite deposits, which is why mineable phosphate reserves are so scarce. Curiously, most older deposits in the world formed during the Ediacaran and Cambrian Periods. In fact, more phosphate is produced today from deposits aged between about 600 and 500 million years old than from all the other periods in earth history combined. This astonishing fact has furnished many a hypothesis to explain the Cambrian radiation of animals by invoking exceptional rates of weathering and a resulting nutrient feast.

The importance of nutrient availability was first brought home to me back in 1997, on my return to Nouakchott after fieldwork in the Sahara Desert. I had several days to kill before my return flight, so I spent that short time getting to know a little about Mauritania's dusty but growing capital city. Nouakchott is an amazing melting pot, where Saharan and sub-Saharan cultures merge, and where desert sands of the western Sahara melt seamlessly into the beach dunes of the eastern Atlantic seaboard. Rather less harmoniously, it is also where the colorful southern fisherfolk mingle with the northern African trader clans. Since independence from French rule in 1960, Nouakchott's population has increased a hundredfold, frequently stretching food supplies to the breaking point. Although staple foods may be hard to grow in the Mauritanian climate, fish, at least, are plentiful. The waters here are among the richest fisheries in the world, all because of the upwelling of nutrient phosphate just beyond the ocean's horizon.

In the cooler evenings, I watched the fishermen, many of whom hailed from neighboring Senegal and Gambia, launch their joyously decorated boats amid much singing and stamping of feet. My reason for being there was rather less exhilarating: I needed a volunteer to get me a sample of Atlantic seawater to act as our laboratory's calcium isotope standard back in Strasbourg. "Dolle, Dolle, Dolle," they shouted, "we are the force," as they paddled through the dangerous swell. The fish rise to the surface at night, making them easier to find, but catching them is still a risky business in small wooden boats that offer little protection against fierce winds and crashing waves. I remember fondly how the fishermen were sung back to shore the following day by excited lady folk, awaiting the haul in eye-catching dresses that often hid tiny children in their folds. The younger returnees, gleaming with sweat and pride, were eager to show the best of their catch. Fish is not just a business in Nouakchott. It is also the language of love.

The local fish markets, though bustling, do not show the

ocean's true bounty. Already in 1997, you could just about glimpse a long line of gigantic container ships far out to sea. Mauritanian waters have long been "rented out" to the highest international bidder. These days, more than a million tons of fish are caught in Mauritanian waters every year, but a mere 5% will be processed locally. As phosphate wells up from the ocean deeps, it fuels a chain reaction of consumption, as the primary producers, or phytoplankton, are eaten by zooplankton that are then eaten by shrimp and in their turn by little fish, bigger fish, and even whales. As each trophic level is attained, phosphate builds up in excrement and bones, which fall through the water column after death, destined to be deposited over the narrow continental shelves.

I have often wondered whether China, during the Ediacaran-Cambrian transition, was as productive as Mauritania today. Phosphate fuels life, and, although fish had not yet evolved, phosphate deposition occurred in southern China on a similarly massive scale over an interval spanning almost seventy million years. Phosphate has been deposited off African shores for about the same amount of time since the Cretaceous Period. Although the Yangtze craton hosts the largest deposits, phosphorite of the same age has also been reported from many other parts of the world. The earliest of these phosphorite deposits formed directly after Marinoan deglaciation across northwest Africa and then Brazil, China, and Mongolia. By the start of the Cambrian Period, phosphogenesis extended right across the globe. When it began, guts had not yet evolved to make excrement, and there were also no bones to parcel up the phosphate like today. Something tells me that the Ediacaran-Cambrian phosphorite giant episode was a very special case in earth history. Fossils may provide a clue.

The oldest animal fossils visible to the naked eye were described in chapter 7. However, the rooted fronds and fractal quilts on the late Ediacaran seafloor might not in fact be the oldest animal fossils. In 1998, the same year that Paul Hoffman injected his tremendous energy into promoting the Snowball Earth hypothesis, a similarly catalytic paper appeared in the journal *Nature*. A team of US and Chinese paleontologists had been studying rocks in the phosphorite quarries of Guizhou Province in southern China. Near a city called Weng'an, they had come across a curious 600-million-year-old rock that was composed of myriad tiny balls, much like an oolite. However, their investigations showed that these were not calcitic and smooth, as with ooids, but phosphatic and ornamented, brain-like spheres, comprising differing numbers of chambers just like embryos in various stages of growth. Shuhai Xiao and his supervisors Yun Zhang and Andy Knoll made the spectacular claim that although some of these were likely to be algal embryos, and thus intriguing but unsurprising, others seemed to show the characteristic cleavage patterns of animals, and possibly even bilaterians. The oldest known fossils of adult bilaterians had only ever been confirmed from rocks fifty million years younger.[3]

Shortly after the discovery, feverish attempts began among paleontologists to extract and decipher these amazing microfossils, but with extremely varied results. While some researchers concurred that these were likely to have been animals, others thought they might have been algae, and still others giant bacteria. Some embryos were even found within the large spiny acritarchs that are typical of early Ediacaran rocks, implying that these "acanthomorphic" acritarchs were resting cysts, or hard cases where a type of multicellular organism was able to sit out difficult times before emerging when conditions improved. Acanthomorphic acritarchs of the global Pertatataka Assemblage have even been discovered in rocks directly above cap dolostones in parts of China, so whatever these spiny cysts were, it seems certain that they were already on the scene by 631 million years ago, only shortly after the earth's icy cover had melted away. It seems highly probable that these were close ancestors of those Snowball Earth survivors that won the evolutionary arms race in the aftermath of catastrophic deglaciation.[4]

Following two decades of forensic detective work and state-of-the-art 3D imaging, consensus opinions have been shifting somewhat over the biological affinities of these enigmatic embryos, away from the extremes of bacteria or bilaterian animals toward a middle interpretation. Although it is still possible that some were indeed the fossilized embryos of early animals, probably before most modern animal phyla had branched off the tree of life, and notwithstanding some convincing candidates for fossil sponges at Weng'an, many, if not all, can be interpreted more plausibly to be the remains of non-metazoan holozoans (Figure 21). The holozoa form a clade, which covers all organisms more closely related to animals than to fungi. We have already heard about the best-known group of non-metazoan holozoans, the choanoflagellates, long thought to be ancestral to sponges and all modern animals. The extraordinary Weng'an fossils have given us a rare window to glimpse some of the very early evolutionary stages on the path to all modern animals. But how did this "taphonomic," or preservation, window come about? The key was phosphogenesis.[5]

It is an amazing fact that a substantial proportion of the world's phosphorus fertilizer is produced from the mining and subsequent processing of fossil embryos of our earliest ancestors. Yet those organisms had no hard parts, and certainly no parts that were originally made of phosphate. So, how did they end up preserved as mineral phosphate? Some are so exquisitely mineralized that their internal structures, including cell nuclei, can still be seen using noninvasive microtomography. Although phosphate mineralization is also known from later

Figure 21. Scanning electron microscope images of typical 600-million-year-old Weng'an fossil embryos. Photos (a) through (c) show specimens of *Tianzhushania* at various stages of division from a single cell to many cells. Photos (d) through (i) represent more specific categories of fossil: (d) *Helicoforamina;* (e) *Spiralicellula;* (f) *Caveasphaera;* (g) *Archaeophycus,* a putative red alga; (h) *Mengeosphaera,* a spiny acritarch or, likely, a resting cyst similar to those that have been found to contain fossil embryos; and (i) *Eocyathispongia,* a possible sponge. Fossils range from about 0.5 to 1 millimeter across. (Photos courtesy of Phil Donoghue, University of Bristol, first published together in Cunningham, J. A., Vargas, K., Yin, Z., Bengtson, S., Donoghue, P. C. J., "The Weng'an Biota (Doushantuo Formation): An Ediacaran window on soft-bodied and multicellular microorganisms," *Journal of the Geological Society* 174 [2017]: 793–802)

times, nothing matches the evident perfection of Ediacaran-Cambrian phosphogenesis. If any decomposition had occurred after death, then none of these fine features would still be detectable, so lithification, or in this case phosphogenesis, must have occurred imminently after death. In other words, the seafloor must have been oversaturated with respect to calcium phosphate. What could have led to such high phosphorus contents? It seems likely that a number of different factors were at play.[6]

Given that all phosphate derives ultimately from chemical weathering, we might justifiably expect times of high weathering rates, such as in the aftermath of Snowball Earth, to coincide with widespread phosphogenesis. There is some evidence for such a connection, but, as pointed out earlier, the vast majority of nutrient phosphorus comes from recycling, so changes to the P sink must have had an even greater role to play, most likely through the sulfur cycle. The major sink for phosphorus is in the form of the mineral francolite, which precipitates within the sediment at or close to a redox boundary where iron, carbon, and sulfur engage in microbially catalyzed cycles of oxidation and reduction. Iron oxides and organic matter are both known to contain phosphorus, which is then released and precipitated as francolite close to this redox boundary. Sulfide-oxidizing bacteria also form part of the story. As they store polyphosphate under oxic conditions, these microbes play a key role in the precipitation of francolite by exploiting sedimentary redox gradients. This is consistent with the observation that phosphogenesis commonly occurs beneath an oxygenated water column, but in China, as elsewhere, it occurs in close proximity with anoxic and even sulfidic waters. It is even thought that sulfidic or euxinic waters may be extremely important for the efficient recycling of phosphorus because, under such conditions, neither the francolite nor the oxide phosphorus sink would be effective. Such was the case at Weng'an, where euxinic waters were in close proximity, as outlined in chapter 8, yet the geochemistry of the embryos themselves shows that the overlying waters contained oxygen, sulfate, and nitrate in abundance. It seems therefore that the giant phosphorite deposits of the Ediacaran and Cambrian Periods were another consequence of a changeable and heterogeneous ocean redox environment.[7]

Oxygen has another consequence for the phosphorus cycle in that organic decay beneath an oxygenated water column generates acidity, which suppresses calcium carbonate precipitation, thereby pushing the calcium phosphate saturation state even higher. When animals began burrowing into the sediment, by about 550 million years ago, bioturbation introduced oxygen and other electron acceptors into the sediment, further encouraging the global phosphorus sink. Oxygenation and the evolution of muscular animals may have even caused a negative feedback on atmospheric oxygen by more efficiently removing phosphate from the ocean, thus limiting productivity, organic burial, and further oxygenation. It seems likely that a perfect combination of free oxygen in overlying waters, little or no bioturbation, and widespread euxinia in highly productive areas like zones of upwelling would have sharpened the zone of phosphatization to the topmost surface of the sediment, allowing crystal apatite to mimic the finest ornamentations on, and even within, individual holozoan cells.[8]

This may explain the locally abundant Weng'an fossils, but such conditions were not unique in earth history. Exquisitely phosphatized microbial laminae and organic tests have been found, although to a lesser extent, at other times of sulfate enrichment and ocean anoxia during the Proterozoic Eon, and may go some way to explaining why calcium phosphate was one of the earliest skeletal biominerals of both the Tonian and Cambrian bioradiations. To build their skeletons, organisms appear to adopt the biomineral that requires the least energy to manufacture, and so is at or close to supersaturation, at the time of those organisms' evolution. Thereafter, they tend to stick to those minerals even after environmental conditions become less favorable. It is intriguing to note that some of the first skeletonized fossils were the phosphatic grasping spines, or *protoconodonts,* of the chaetognaths. Also known as arrow worms, these transparent, torpedo-shaped creatures were among the earliest bilaterian carnivores. The phosphatic legacy of our distant past may even help to explain why we build our own bones, somewhat bizarrely, from calcium phosphate.[9]

Let me sum up this complex picture. There is good evidence to suggest that anoxia, and specifically euxinia, leads to more efficient phosphorus recycling. This represents a positive feedback because nutrients fertilize organic production, which in turn exacerbates anoxia when that organic matter starts to decay. However, the ultimate result of runaway eutrophication is the release of oxygen due to organic carbon and/or pyrite sulfide burial, both of which are tied to primary production. Therefore, at some point in the process, positive feedbacks toward greater regional anoxia give way to a longer-term negative feedback, just as soon as enough oxygen is released into the atmosphere to tame the runaway train. A complementary negative feedback might operate through the effect that organic carbon burial has on cooling the climate, thus decreasing the rates of chemical weathering and nutrient phosphate flux. Another could operate through the sulfur cycle, whereby pyrite burial lowers seawater sulfate concentrations, inhibiting rates of sulfate reduction. There is also a fourth possibility, however, that no negative feedbacks were robust enough during the Ediacaran-Cambrian transition to prevent destabilizing positive feedbacks from dominating. The extreme amplitudes of carbon cycle perturbations, in the form of unprecedented carbon, sulfur, and uranium isotope and climatic events, seem to be telling us that regulatory feedbacks were so weak as to be elastic. But what caused the biosphere to lose its resistance to change? One key player in this loss of resilience must have been tectonics, which is not only the biggest player in the earth system, but also the least susceptible to surface feedbacks. So, what was happening tectonically in the Ediacaran world, and were those changes consistent with an accelerated nutrient flux?

The first and only time I visited the celebrated Grand Canyon was in September 2019. After both speaking at the Geological Society of America, or GSA, conference in Phoenix, Ying and I rented a gas-guzzling monster of a truck with a six-liter engine—we had asked for a modest sedan—with the only plan being to gaze in awe at possibly the most famous geological spectacle in the world. My first reaction was surprise at how familiar the Grand Canyon seemed to me. The view looking across to the northern side is, after all, pictured in any number of introductory geology textbooks, and I have shown the same view myself in lectures for the best part of twenty years. The most striking aspect of the canyon, for a geologist, is the angular unconformity that separates Precambrian from Cambrian rocks. Christened the Great Unconformity, its existence supports the long-held view that a sedimentary break occurred beneath the Cambrian System in many parts of the world. Called the Lipalian interval, this hiatus was thought to be responsible, at least in part, for Darwin's Dilemma: the seemingly abrupt appearance of animal fossils in the rock record. We no longer think of a global Lipalian interval as such, but it is striking just how many places in the world have unconformities of this approximate age. Clearly, the Ediacaran-Cambrian transition interval was a time of substantial tectonic uplift and erosion. Much of the break in the Grand Canyon can be attributed to the Cryogenian and/or mid-Ediacaran glacial scouring millions of years earlier. As we shall see, however, this cannot explain all the erosion. The Ediacaran Period, in particular, witnessed a "clash of the titans" as numerous smaller blocks collided to form continents in one of the greatest mountain-building episodes, or orogenies, ever.[10]

High sedimentation rates and the sheer volume of sandstones around the world support the notion that the Ediacaran-Cambrian transition witnessed exceptional amounts of uplift and erosion. There is plenty of geochemical evidence, too, for such upheaval. The strontium isotope ratio of seawater—we encountered that global weathering tracer in chapter 5—shows its greatest known rise during this interval, providing convincing evidence for the uplift and erosion of old cratonic interiors, presumably as a result of tectonic collisions. We can also trace this erosion by looking at the mineral contents of the sediments themselves, the most useful being zircon.[11]

As we saw earlier, zircons are the best radioactive clock for measuring the ages of rocks. To obtain suitably precise U-Pb ages, tiny magmatic zircons need to be separated out from rocks by a long and exhaustive series of techniques, beginning with crushing, sieving, and mineral separation using magnetic fields and heavy liquids, such as toxic bromoform, methyl iodide, or the far less lethal sodium tungstate. Zircon crystals are then picked using an ultrafine needle under the binocular microscope, investigated for purity using cathodoluminescence, abraded by pressurized air, and etched with acid before being entirely dissolved, evaporated, dissolved again, passed through cation exchange columns to concentrate uranium and lead, and finally analyzed on the mass spectrometer. And all that without contaminating the samples with even the tiniest picogram amounts of lead from the environment. In Zurich, I once prepared three rock samples for dating and picked them for days on end under the microscope. After that, every time I closed my eyes for weeks I saw only zircons, and even dreamed of zircon people. The Zurich laboratory was among the cleanest, for lead, in the world, and the lead isotope work was all carried out in an innermost sanctum: a positively pressured laboratory inside the already positively pressured clean lab, which itself was inside the wider geochemistry lab. My fellow doctoral students and I found it amusing that, despite the extreme measures taken to keep the place extraordinarily free of contamination, every summer without fail a steady stream of tiny ants would come out of the taps. The lead they brought into the lab did not seem to contaminate it, so they were generally ignored. It is an exhaustive and extremely technically challenging thing to date a single zircon crystal using high-precision Thermal Ionization Mass Spectrometry, or TIMS for short, but absolutely essential if age errors of ± 0.05% are to be obtained. Fortunately, there is another way of dating zircons, by zapping them with a laser. This approach sacrifices precision, but makes up for it with the speed with which huge numbers of ages and other geochemical data can be produced.

Over the past two decades, technological advances in mass spectrometry, namely Laser Ablation Inductively Coupled Plasma Source Mass Spectrometry, or LA-ICP-MS, have revolutionized the field of zircon dating and with it our understanding of earth history. It is now possible to take the zircon mineral separates and date them directly, simply by aiming a laser, with a single session producing tens or even hundreds of ages. Because the ages of sedimentary zircons represent the catchment area undergoing weathering, age distributions provide vital clues about the rocks being weathered and the tectonic history of the area. The total number of zircons dated this way now is fast approaching one million, and a global picture of magmatic zircon abundance through time has emerged, showing well-defined peaks and troughs of zircon formation,

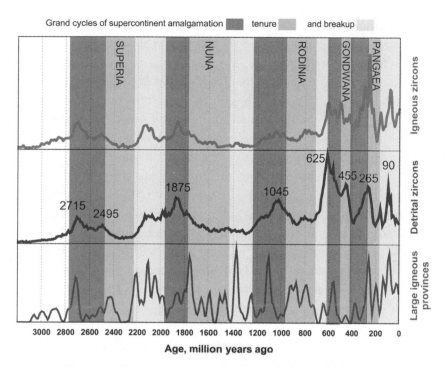

Figure 22. Supercontinent cycles through the earth's history. Tectono-magmatic or supercontinent cycles through the lens of the zircon abundance and magmatic activity records. Each abundance acme represents orogenic, or mountain-building, peaks related to the amalgamation of a supercontinent, while abundance zeniths reflect tectonic lulls during their prolonged erosion, denudation, and eventual breakup.

likely related to the eventful amalgamation and quiescent ten-ure phases of supercontinents, respectively (Figure 22). Specu-lative notions that there were Eurasia-sized megacontinents or even supercontinents before Pangaea, such as Superia, Nuna, and Rodinia, are today based not only on geological interpre-tations but on the abundance, distribution, and isotopic com-position of zircons. One further advantage of laser ablation is that the same zircons can also be analyzed for a range of other isotopic systems, providing yet more clues about our planet's tectonic history.[12]

The zircon abundance record shows high peaks at 1.85 and 1.05 billion years ago, which relate to tremendous orogenies associated with the amalgamation of Nuna and Rodinia, respectively. The "boring billion" is confirmed as a time of relative stability. The next peak, centered around 600 million years ago, falls squarely within the Ediacaran Period, when strontium isotopes provide independent evidence for uplift and erosion. Intriguingly, the Ediacaran-Cambrian collisions are characterized by uniquely extreme mean values in both hafnium and oxygen isotopes in detrital zircons globally. The hafnium isotope composition of zircons corroborates the strontium isotope story that orogenies must have involved the uplift and erosion of ancient terranes, while the oxygen isotopes tell a tale of unprecedented amounts of crustal reworking. But how? What on earth was happening to cause this unprecedented tectonic acme?[13]

In today's world, such immense uplift, erosion, and reworking are mostly associated with the Himalayan mountain range, which marks the place where India and Asia collided fifty million years ago. These two crustal blocks are now welded together along a suture zone 2,400 kilometers in length stretching from Afghanistan to Myanmar. The mountain chain reaches such heights because the ocean that once existed between the blocks has been subducted into the earth's mantle. The squeezing and melting of the thickened continental crust caused the low-density crust to bob up like an ice cube in water, exposing once deeply buried and ancient rocks to the elements at high altitudes where they could be weathered more rapidly. In 2010, two Australian geologists, Ian Campbell and Richard Squire, estimated the Ediacaran-Cambrian orogeny to have stretched across more than 8,000 kilometers, more than three times the size of the Himalayas today. They named the resulting mountain chain the Transgondwanan Supermountain, the erosion of which produced more than 100 million cubic kilometers of sediment, enough to bury the United States with a layer of rocks ten kilometers thick. The "supermountain" marked the suture between two colliding megacontinents: West Gondwana, comprising Africa, which at that time was still joined to South America, and East Gondwana, which was an amalgamation of Antarctica, India, Australia, and Arabia.[14]

According to Campbell and Squire, weathering of the supermountain would have brought tremendous quantities of sediment into the oceans, burying organic carbon, releasing oxygen, and thereby paving the way for animal diversification. Nutrient phosphate, which can only come from chemical weathering, fueled the organic production, and other products of weathering, such as calcium ions, led to biomineralization. Many such "weathering theories" for the Cambrian radiations have come and gone, and for good reason. Organic burial results in positive carbon isotope anomalies, but the Ediacaran-Cambrian transition interval—as we shall see in the next chapter—is characterized, almost uniquely in earth history, by negative anomalies. Uplift and erosion not only help to produce oxygen via sedimentation and nutrients, but also expose reduced forms of carbon, sulfide, and iron to weathering, and so are tremendous consumers of oxygen, too. It is not immediately clear why acceleration of the rock cycle by tectonic uplift alone would necessarily lead to any net surplus in the earth's oxygen budget. Finally, because oceans are saturated with respect to calcium carbonate, greater weathering would increase calcium carbonate deposition, but would not necessarily lead to a higher saturation state, thus tipping the balance toward biomineralization. The impressive tectonic upheavals of these times are clearly important to the story of our origins, and they would indeed have caused nutrient phosphate delivery to the oceans well in excess of more normal times, but there has to be more to it than simply moving material from mountain tops to the seafloor. We will keep our powder dry for the moment, until the next chapter, when these various threads finally come together. In the meantime, let's complete our look at nutrients by briefly examining life's other critical nutrient element: nitrogen.

Although nitrogen is readily available in the atmosphere, we should not make the same mistake as Justus von Liebig in thinking that it cannot be a limiting nutrient just because it is plentiful. Only a very few organisms can utilize nitrogen in its elemental form. All such nitrogen-fixing, or diazotrophic, organisms are either bacteria or archaea, while other organisms, including all eukaryotes, need to obtain their nitrogen either, like legumes, through symbiosis with diazotrophs, or in already fixed forms, such as nitrite, nitrate, or ammonia. This poses problems for photosynthetic algae in a largely anoxic ocean because they can be outcompeted easily by cyanobacteria. Fixed nitrogen is lost during a process called denitrification, which returns it back to its unavailable elemental state. Today, most denitrification takes place within the sediment because the seafloor is generally oxygenated, so upwelling currents carry nitrate as well as phosphate. However, in the anoxic oceans of the past, denitrification would have taken place more commonly within the water column, starving algae of fixed nitrogen and drastically reducing the areas of the ocean where they could survive. Until the oceans became more widely oxygenated, it seems unlikely that eukaryotic algae could have unshackled themselves from their dependence on diazotrophic microbes. This could explain why algal biomarkers only became significant during the Neoproterozoic Era.

Nitrogen fixation happens anaerobically, so cyanobacteria have evolved ways to separate the nitrogen fixation mechanism from today's highly oxygenated environment. Most cyanobacteria fix nitrogen using heterocysts, which can be identified as slightly enlarged cells in the string of beads typical of coccoid cyanobacteria. When this ability first evolved is still unclear, but enlarged cells have been found in fossil cyanobacteria from North China that date back to early Tonian times. Intriguingly, some molecular studies have highlighted the Tonian Period as a time of significant diversification among cyanobacteria, marking the moment when open ocean nitrogen fixation became possible. Why both of these biological innovations would have come so late in the day is still a mystery, although the coincidence in timing between these events and the emergence of photosynthetic red and green algae suggests that much more on the topic of early Tonian biodiversification will be discovered in due course. I would argue that oxygen, once again, is key to early eukaryotic radiations, as nitrate recycling would permit eukaryotes to compete with cyanobacteria, which in turn needed to evolve protection against the added oxygen. The first fossils that can be ascribed to the typically Ediacaran discoid fossil *Aspidella* have also been found in early Tonian rocks on the North China craton. Discs seem always to appear at such pivotal moments; we learned earlier of similarly cryptic appearances after the Sturtian ice age in Canada and, of course, following the Gaskiers glaciation in Newfoundland. Perhaps the early Tonian period witnessed an initial ocean oxygenation event, a precursor to the full-blown marine biodiversification to come.[15]

One might suppose, then, that organic production would be limited by fixed nitrogen availability in anoxic oceans and by phosphorus availability in oxic oceans. However, because cyanobacteria can fix nitrogen independently, it might be more appropriate to consider that anoxic oceans favor a microbial eco-system with varying populations of eukaryotes at the periphery, while in more oxic oceans the balance shifts toward eukary-otes. Our research group has enlisted nitrogen isotopes to dem-onstrate this balancing act through the Ediacaran-Cambrian transition, as denitrification leaves a characteristic isotopic fingerprint. The data so far suggest that the Ediacaran ocean contained a stable nitrate pool, a bit like in the modern ocean, but that this pool waxed and waned throughout the Ediacaran and Cambrian Periods, in line with changes in other oxyanions like sulfate, molybdate, and vanadate. It seems probable not only that animals and other obligate aerobes enjoyed success whenever conditions were favorable, but that they were joined by other forms of eukaryotic organisms, like algae and protists, too. With each expansion in habitable environment, new forms of opportunistic aerobes radiated around the world.[16]

The oxygenating ocean clearly had a great effect on bio-geochemical cycling. The eventual clearing of ocean turbidity by oxidation and the actions of animals would have made algae and other eukaryotes even less dependent on the hand-me-down fixed nitrogen from microbes. Although Sprengel's lim-iting nutrient, for eukaryotes at least, likely shifted between phosphate and nitrate at times, super-greenhouse conditions and supermountain orogenies after Snowball Earth were deliv-ering phosphate in great abundance to the world's oceans. Eu-xinic conditions in productive marginal seas, such as in China, were ideal for phosphate recycling, helping to drive positive feedbacks toward runaway organic production, pyrite burial, and oxygen flux. Eutrophication, therefore, produced sharp redox gradients in the marine environment, with oxic condi-tions neighboring anoxia, either on the shallow shelves or in the topmost layer of the microbial carpet seafloor. Widespread early phosphatization was the result, and, intriguingly, Protero-

zoic phosphorite from other times of transient ocean oxygenation, at about 2.0 or 1.6 billion years ago, was similarly early.
The redox boundary, and with it the phosphogenesis window,
moved deeper and later once bioturbating animals evolved and
oceans became more consistently oxygenated. We have the oxygenating ocean of the Ediacaran-Cambrian transition to thank
for much of our food security and fertilizer supply today.[17]

Much of what we have learned about the earth system so
far suggests that unstoppable unidirectional change in nutrients, climate, or oxygen is not the way the world is supposed to
work. Negative feedbacks are usually taken for granted to imply
that, for example, what tectonics gives with one hand, other
processes take away, stabilizing the status quo. Many elemental
cycles played a role in the emergence of animals and the modern earth system. There were important roles for phosphorus
and nitrogen and, as we know, a starring role for oxygen. Climate change, tectonics, and weathering all had their parts to
play, too. But what we have not discovered yet is who the rogue
players are in this game of perpetual cycling. If everything is
always in near perfect balance, what can drive seriously major
disruptions like Snowball Earth? How did the Cryogenian ice
ages happen? What caused the silicate weathering climate feedback to weaken? What caused the extreme fluctuations of oxygen during the Ediacaran-Cambrian transition that permitted early animals to radiate ever more widely with each oxygen
boost? Now that we have a better understanding of the roles of
nutrients like phosphate and nitrate, we can begin to see how
these ingredients came together to turn the earth system into
a bobbing boat on a rough sea, with precious little ability to
right itself before the next wave came in. But we are not quite
there yet. If we can only work out how those pesky carbon cycle
fluctuations came about, maybe we will have the final piece of
the puzzle. After all, it can hardly be a coincidence that carbon

isotope excursions, more than anything else, characterize this entire interval of transition from the Tonian through the Cryo-genian and Ediacaran to early Cambrian times. As I intimated earlier in this chapter, tectonics is key, as are nutrients, but if we want to understand how, we need to get to grips with the elephant in the room: negative carbon isotope anomalies.

10

A Pinch of Salt

From before Snowball Earth through the Cambrian explosion (750–500 million years ago) the global carbon cycle experienced instability, evidenced not only in extremes of climate, but also in wild carbon isotope fluctuations. The geochemical record reveals a largely anoxic ocean with vast amounts of organic matter that waxed and waned in response to cycles of bacterial sulfate reduction linked to the weathering of vast salt deposits on land. Recent new evidence brings together all the extraordinary events of this time to show how abrupt radiations of animal life forms (Darwin's Dilemma) were linked to both orogeny and oxygen.

It was thirty years ago, though it hardly feels like that long, that I first arrived in Zurich, more blotchy than bronzed, after a few months of selling snacks on the banks of Lake Lucerne in central Switzerland. For an entire summer I had sauntered each morning the few miles to my *Bademeister*'s cabin near Meggen, only to saunter lazily back again when evening came around. This was a time when all my worldly belongings could squeeze into a small duffle bag. My only constant companions were a few clothes, the odd trusty book or two, and a glossy magazine called *Erdwissenschaften heute,* or *Earth Sciences Today.* The magazine described the exciting scientific discoveries being made at the ETH, short for the Eidgenössische Technische Hochschule, in Zurich, which is where I was planning to spend a year studying a mixture of geology and German before starting a PhD. Thumbing through it once more on the train, I was intrigued by the work of one group in particular there, led by Professor Ken Hsu, who, although I did not know it at the time, was quite possibly the most famous geologist in the world.

Hsu had enjoyed a stellar career in geology since leaving China for the United States in 1948, and his fame was due in no small measure to the popular science books he had written along the way. His best-selling reads had picture-book titles like *The Great Dying,* about how a meteorite extinguished the flightless dinosaurs, and *The Mediterranean Was a Desert,* about the discovery of huge amounts of salt beneath the Mediterranean seafloor. He had also come up with a number of catchy metaphors, like the "Strangelove ocean" to describe the eerie aftermath of mass extinctions. My arrival at the ETH proved fortuitous, as Hsu was starting to turn his attention to the Cambrian explosion, inspired in part by James Lovelock's Gaia hypothesis, which postulates that life begets conditions favorable to its own success. We will return to Gaia at the end of the book, but it is Hsu's Mediterranean work that interests us now. Over the intervening three decades between then and now, seemingly disparate clues, namely glaciation, nutrients, oxygen, and tectonics, have gradually merged to form a more holistic explanation of our evolutionary origins. Little did I suspect back then that the final missing ingredient would be salt.

A generation earlier, Ken Hsu and Bill Ryan had been joint leaders of a groundbreaking drilling cruise aboard the research ship *Glomar Challenger*. During that cruise, they discovered thick salt deposits beneath the seafloor in some of the deepest parts of the Mediterranean Sea, which averages 1,500 meters but can be as deep as 5,000 meters. Looking for salt had not been their original intention. The Mediterranean Sea was originally the western end of a much larger ocean basin called the Tethys Sea, named after the Titan goddess who was daughter to Gaia and wife to Oceanus. Leg 13 of the international Deep Sea Drilling Project was supposed to find out when and how Tethys had closed. Hsu and Ryan hoped to confirm that the Tethys Ocean lithosphere, which is the brittle top portion of the solid earth comprising the crust and uppermost mantle, was sliding, or subducting, northward beneath the volcanoes of southern Europe, crumpling the crust and ramping up the Alps in the process. The year was 1970 and tectonics was still not universally accepted, so finding evidence for subduction in the deepest trenches of the Mediterranean Sea was an exciting prospect that might help to further validate plate tectonic theory. The discovery of salt under its deepest parts now raised the question of whether, at some point during the relatively recent geological past, the entire Mediterranean Sea had evaporated away entirely, leaving behind only salt.

By finding salt in the Mediterranean Sea, Leg 13 confirmed earlier suspicions that its western gateway into the Atlantic Ocean had become severely restricted near the straits of Gibraltar toward the end of the Miocene Epoch, only a little more than five million years ago. The name Messinian Salinity Crisis had been proposed for this event just a few years earlier based on thick layers of gypsum, a form of hydrated calcium sulfate, at Messina in Sicily and elsewhere close to the margins of the Mediterranean Sea. Gypsum is a soft mineral, carved through the ages to make alabaster sculptures all the way from Mesopotamia to Spain. Miocene alabaster is still mined today around the Mediterranean region. The idea that the Mediterranean Sea was once dry is not new and has been a frequent topic of speculation through the ages, from the writings of the Roman scholar Pliny the Elder to the science fiction of H. G. Wells. Inundation by the Atlantic Ocean may have even inspired Plato's Atlantis myth. However, once it was discovered quite how deep the Mediterranean Sea is—it is, after all, the remnant of an ocean—such fanciful notions subsided. The extraordinary extent of the Messinian Salinity Crisis remained entirely unsuspected until the drilling cruise of 1970.

The first minerals to fall out of solution when seawater evaporates are carbonate minerals, like the limescale in your kettle. As seawater evaporates further, gypsum will begin to precipitate once salinity has risen more than threefold. However, this threshold depends very much on the concentrations of other ions in solution, namely sodium and chloride. In the modern Caspian Sea, for example, calcium sulfate minerals will precipitate already at only just above normal ocean salinity because of the relative lack of ions other than calcium and sulfate. Some researchers have speculated that the Caspian Sea and the Black Sea, which are also remnants of the Tethys Ocean, supplied the Mediterranean Sea from the east with low salinity, gypsum-saturated waters during the Messinian Salinity Crisis. By contrast, rock salt, or halite (NaCl), is much more soluble and does not begin to precipitate until very extreme conditions are reached after seawater has been evaporated to more than ten times its usual salinity. In the final stages of evaporation, several even more soluble salts fall out of solution. The rims of the Mediterranean Sea are draped by hundreds of meters of gypsum, forming the most obvious signs of the crisis. However, beneath the greatest seafloor depths, halite deposits more than three kilometers thick dominate. The existence of this salt layer beneath the deep seafloor had been discovered before 1970, when seismologists noticed a persistent reflective layer. However, the offshore salt layer had always been presumed to be much older than the onshore gypsum deposits, possibly relating, it was thought, to very early stages of seafloor spreading. The young Miocene age of the offshore salt layer became clear only when it was drilled and was a complete surprise to Hsu and the rest of the science crew of the *Glomar Challenger*.[1]

Once geologists were able to link the onshore and offshore salt deposits, the enormity of the Messinian Salinity Crisis astounded the wider scientific community, and fierce arguments ensued. Despite the catchy title of Ken Hsu's book, it quickly transpired that the possibility of the entire Mediterranean Sea becoming a desert for much, if any, of this time was extremely unlikely. Such thick evaporite deposits cannot be the result of desiccation alone but require the repeated replenishment of seawater. Evidently, the amount of salt, both halite and gypsum, was far more than all of the dissolved salt in the Mediterranean Sea today. Either the Mediterranean Sea dried out many times during the Messinian Age or, more likely, it remained partially open to the Atlantic Ocean or the central Tethys (Paratethys) Ocean, or both, for most of that time. In hydrology, this partial evaporation is called *negative water balance,* which occurs when the rate of evaporation exceeds the rate of replenishment through seaways, rain, and runoff. The Mediterranean Sea still has a small negative water balance today and with it significantly higher salinity than the Atlantic Ocean. For an interval between 5.9 and 5.33 million years ago, negative water balance must have become unusually severe, leading to the deposition of more than a million cubic kilometers of evaporite minerals. Events like the Messinian Salinity Crisis are so huge that they can even lower global ocean salinity.

Such *evaporite giants* are rare occurrences related to basin restriction. They tend to happen in very warm parts of the world where ocean basins are beginning to open, but they can also occur when they are just about to close, as happened during the Miocene restriction of the Tethys Ocean. As a consequence, their distribution in time and space is the fortuitous product of climatic and tectonic happenstance. Salinity crises of at least equal magnitude to the Messinian event occurred during the Tonian Period, the Ediacaran-Cambrian transition, and the late Devonian, late Permian, middle Jurassic, and early Cretaceous Periods, averaging one giant event only every hundred million years or so. The Cretaceous Period event was caused by the zipper-like opening of the southern Atlantic Ocean, when seawater spilled into rift basins that now lie deeply submerged off the eastern Brazilian and western African seaboards. Like the Miocene deposits of the Mediterranean Sea, most of these Cretaceous evaporites are still buried far beneath the seafloor. They will only see the light of day once those now passive continental margins undergo uplift and erosion, most likely after the onset of plate subduction and eventual collision. Because evaporite minerals are so soluble, they weather very easily by dissolving entirely. It has been estimated that huge amounts of evaporite minerals, long buried on the margins of the eastern Tethys Ocean, were washed into the sea when India first crashed into Asia, around fifty million years ago. That event would lead eventually to the rise of the Himalayan mountain chain and the Tibetan Plateau.[2]

I find evaporite deposits fascinating because they represent a potentially painful thorn in the side of earth system regulation. Unlike most other parts of James Hutton's grand rock cycle, evaporite deposition and weathering are seldom in balance even over huge time spans, with gargantuan deposition events superimposed sporadically against a backdrop of continuous weathering. Both deposition and weathering of evaporites are driven by tectonic events, which are less susceptible to the stabilizing negative feedbacks that we presume govern climate and the carbon cycle. In recent years, a role for evaporite dynamics in destabilizing the earth system has come to the fore—after having been overlooked for decades—and could provide answers to some of the questions we have been asking throughout this book, such as: Why was the end of the Proterozoic Eon marked by unprecedented climatic extremes like Snowball Earth? Why was it at this moment in earth history that the deep ocean became oxygenated, while ocean margins turned anoxic and even euxinic? Strange as it may seem, evaporites might even have contributed to biomineralization. To understand how, we need to return to carbon isotopes. Thus far, I have mentioned carbon isotopes only as a backdrop to everything else that was going on in the world during the deep past. We geologists use carbon isotopes so often as a tool for correlation and as a metric for carbon cycle dynamics that we sometimes overlook how little we truly understand them. Nevertheless, carbon isotope fluctuations are direct tracers of perturbations to the global carbon cycle, and the carbon cycle is instrumental in governing both climate and oxygen. If we are to really grasp the nettle, we need to get to the bottom of what drives carbon isotope change, and in particular how to explain the uniquely wacky negative carbon isotope anomalies that persisted for more than 200 million years from the pre-Cryogenian all the way through to early Cambrian times.

Ever since my PhD days, I have struggled over the meaning of carbon isotopes, and negative carbon isotope anomalies in particular. When I asked what topic might make a decent PhD project, Ken Hsu pointed me toward the problem of the Precambrian-Cambrian boundary, and so I spent many months writing an application for funding to study the classic boundary sections of Yunnan, Sichuan, and Hubei Provinces in south-western China. Seven years earlier, in 1985, Hsu had made the first ever report of a negative carbon isotope anomaly from the Ediacaran-Cambrian transition in the journal *Nature*. I eventually spent five weeks in 1992 visiting those same sites on a shoestring, with the help of two students from Beijing who taught me a smattering of basic Chinese, mahjong, and which cigarette brands make the best bribes to get a soft carriage on the long, uncomfortable train journeys. Shishan Zhang, my guide to the classic boundary section at Meishucun in Kunyang, Yunnan Province, almost thirty years ago, is still showing visitors around the quarry and his private fossil "museum." On my last visit to the area in 2019, he again showed me the neighboring Wangjiawan section, and its abundant glacially striated pebbles from the Marinoan ice age.[3]

My supervisor, I soon realized, was quite a celebrity back in his homeland, and I felt privileged to benefit by association as quarries and outcrops were opened up to me that had previously been inaccessible to foreigners. In fact, it was better than that. I happened to mention one day that Professor Hsu was about to retire, making me his last ever doctoral student. After that passing comment, I was suddenly treated with entirely unearned respect as the person who was expected, at least symbolically, to carry on Hsu's great legacy. In his pioneering paper, Hsu had attributed the low $\delta^{13}C$ values to a global marine ecosystem crash and, with characteristic flair, dubbed the eerie aftermath of such an event the "Strangelove ocean," after *Dr. Strangelove,* the classic Cold War film directed by Stanley Kubrick. Hsu reached those conclusions because negative anomalies, which represent a deficit in the heavier ^{13}C isotope relative to the ^{12}C isotope, imply anomalously low levels of organic production on a global scale (Figure 23). Similar anomalies follow both the Cretaceous-Paleogene and Permian-Triassic mass

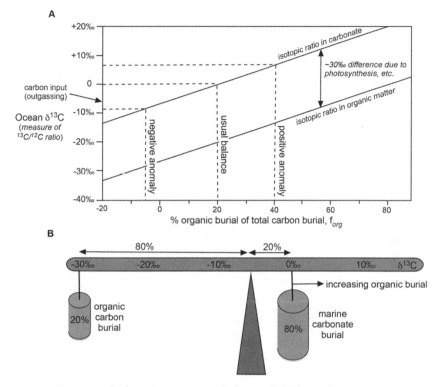

Figure 23. Carbon isotope mass balance. (A) The carbon isotope "Lever Rule" illustrates how the carbon isotope ratio ($\delta^{13}C$) in the ocean depends on the proportion of carbon buried as fossil organic matter. Today's value, zero per mil (0‰), relates to about 20% organic carbon burial, assuming a constant 30‰ difference between deposited carbonate and organic carbon. (B) The lever's fulcrum is placed at the input value (−6‰). Carbonate values lower than this input value imply negative carbon burial, or that more organic matter is oxidized than produced, representing a huge oxygen demand on the earth's exogenic system that can only be supplied by other systems such as the sulfur cycle.

extinctions, so it was reasonable to apply the same logic to the Precambrian-Cambrian boundary interval. When Paul Hoffman and collaborators reported similar negative anomalies from the much older cap dolostones of Namibia in 1998, they initially used precisely the same arguments and interpreted a Strangelove-type productivity crash due to Snowball Earth. There was a problem, however, with these conventional interpretations. Many of the reported carbon isotope ratios were lower than those of any other time in earth history, so much so that they proved increasingly difficult to interpret, especially once even more extremely negative anomalies were discovered over time. Such low values imply that the oxygen released by organic production was less than the oxygen consumed in weathering, resulting in an oxygen deficit right at the time that most geochemists were predicting ocean oxygenation. Understandably, this did not seem to make much sense.[4]

Like many others at the time, I admit, I felt a little skeptical at the start of my PhD that Hsu's findings were representative of the global ocean and occasionally wondered aloud whether the suspiciously large amplitude fluctuations of the Ediacaran-Cambrian transition had been altered somehow after deposition. I was soon dissuaded. Having just returned to Zurich from my trip to China, I was invited to visit Bern, Switzerland's capital, to meet up with a research group led by Professor Albert Matter. His group had been working in the deserts of Oman, and Steve Burns, head honcho in the stable isotope laboratory, generously showed me their labs and even their new, unpublished isotope data. Astonishingly, their data showed even greater negative anomalies than Hsu had discovered, superimposed upon a steep rise in seawater $^{87}Sr/^{86}Sr$. They were able to correlate their negative carbon isotope anomaly, which centered on the late Ediacaran Shuram Formation, across multiple sites, always at the same stratigraphic level.[5]

At the same time that Burns and Matter were pondering the meaning of their Oman data, a group in Sydney, Australia, was mulling theirs. Malcolm Walter, of Macquarie University, had a team of students working on various stable isotope systems, and one of them had discovered an almost identical excursion in the late Ediacaran Wonoka Formation, which overlies strata with large spiny acritarchs, but underlies the world-renowned Ediacaran biota, such as the ribbed pancake fossil *Dickinsonia* or eight-armed *Eoandromeda*, together with evidence for bilaterian-type locomotion. Although debates over the primary versus secondary origin of such extreme carbon isotope anomalies rumble on even today in academic circles, the number of naysayers has dwindled to very few. Since the discovery of the Shuram, or Shuram-Wonoka, anomaly, analogous excursions have been confirmed in correlative strata in China, Russia, the United Kingdom, Canada, the United States, Mongolia, and indeed wherever there are rocks aged between about 575 and 560 million years old. The anomaly consistently appears above the level at which large spiny acritarchs are no longer found and below the appearance of the three "B"s: bilaterians, biomineralization, and bioturbation, or trace fossils. Global consistency increases our confidence that negative anomalies are real, primary phenomena, but it has not made them any easier to interpret.[6]

Hsu's negative anomaly was also eventually confirmed. It is now known to precede the evolution of the trilobites, the poster children of the Cambrian explosion, and it is accompanied by several other isotopic fluctuations, which diminish in amplitude through the early Cambrian. Although the Ediacaran-Cambrian transition interval has undoubtedly the best resolved record of negative excursions, rare anomalies are now known to have occurred sporadically throughout the preceding billion years or more. They precede evidence for large benthic organ-

isms at about 1.6 billion years ago, for example, and both precede and succeed Cryogenian glaciations. Despite their ubiquity, geochemists still find it very hard to come to terms with
these anomalies for the simple reason that they are just too big
to explain away using conventional logic. Although Cambrian
anomalies are somewhat larger than later excursions, even they
are dwarfed by the greatest anomalies, of which there were at
least two in the Ediacaran Period. The Bern group's Shuram
anomaly is still the largest, but there is another very large excursion shortly before the Marinoan glaciation, called the Trezona anomaly after the formation in Australia where it was first
reported. Other smaller anomalies are found scattered throughout the Tonian to Cambrian interval, including just before the
onset of the Sturtian ice age. The carbonate carbon isotope
$^{13}C/^{12}C$ ratios, generally reported as $\delta^{13}C$ values, of the three
largest anomalies are much lower than any other marine car-
bonate values in earth history.[7]

At this point, the nature of the problem with low $\delta^{13}C$
values may seem rather esoteric. However, it is important to
reach a good understanding of carbon isotope anomalies if we
are to solve the puzzle of this extraordinary but key interval in
the earth's past. Carbon isotope anomalies seem to present a
tangible record of the carbon cycle perturbations of this transition, but, rather than helping us to understand what happened,
they have stood in the way because they appear to defy the universal principle of mass balance. As a result, in the thirty-five
years since Hsu's discovery of the first such anomaly, they have
been repeatedly attacked as mistakes, artifacts, or unrepresen-
tative local phenomena. From previous chapters, we are already
familiar with the concept that sources and sinks of an element
into a reservoir like the ocean or atmosphere must equal each
other on certain time frames. For the modern short-term carbon cycle, only decades are needed to balance the atmospheric

carbon dioxide budget. For the long-term geological carbon cycle, however, 100,000 years is needed for all net carbon sources and sinks to match up, as that duration exceeds the residence time of carbon in the largest surface reservoir: the oceans. However, if the isotopic composition of the ocean reservoir remains stable on this same time frame, then it follows that the isotopic composition of sources and sinks must also be in balance. The principle of isotope mass balance requires, then, that over long time periods carbon sources and sinks have precisely the same $\delta^{13}C$ value, which is commonly assumed to be that of the bulk earth (about −5 per mil, or −5‰, relative to the modern ocean).

The four main natural sources of carbon dioxide are volcanic outgassing, metamorphic decarbonation, carbonate precipitation, and oxidative weathering of fossil organic matter. The three main sinks of carbon dioxide today are silicate weathering, carbonate weathering, and organic burial. Because negative anomalies lasted for millions of years, they ought to be explicable using the isotope mass balance principle, and yet their low values, with sustained periods as low as −10‰, demonstrate that the carbon sources and sinks were far richer in ^{12}C than any known source with the exception of organic matter. The puzzling implication of this is that the oxidative flux from organic carbon into surface reservoirs back then must have been very high, and even higher than organic burial. In other words, more oxygen must have been consumed by carbon weathering than could possibly have been released by carbon burial, causing an oxygen imbalance that lasted for millions of years. Although the rise of the Transgondwanan Supermountain exposed more organic matter to weathering, the surplus oxidative flux cannot be due to increased uplift and erosion alone, as the isotopic effect of increased oxidative weathering would be canceled out by increased carbonate weathering. In any case, the major problem here is that there is simply not enough oxygen, even today, to sustain such an imbalance, so how could there possibly have been enough such a long time ago when oceans were widely anoxic?

Extraordinary data require extraordinary explanations, yet the conventional explanation of negative anomalies from standard carbon isotope mass balance was so extraordinary that few believed it possible. When the Bern group tried to convince fellow geochemists in the early 1990s that the anomaly was genuine, their skeptical peers proved too difficult to sway, so the authors eventually gave up pushing their original interpretation. Subsequent researchers have met the same intransigence, and even today may fail to overcome the skepticism of their peers. For those unfamiliar with the peer review process, all published papers are generally reviewed by two or more "independent" experts. A dedicated subject editor will then accept the paper without any changes (unusual), or recommend revisions based on the expert reviews. In high-impact journals, the most common journal response is simply rejection of the paper either before or after review. Reading between the lines of the rejection letter can either encourage authors to send back a revised version for further consideration or give up and try elsewhere. Just in 2020, a paper by a junior colleague received one of the meanest and most scathing rejections I have ever read: "The . . . enormous swings in atmospheric O_2 [mean] that the whole hypothesis [makes] no sense." "This paper should be rejected outright and I discourage the authors from resubmitting. The entire foundation of this paper is deeply flawed, and I would recommend that the project be discarded." Harsh, even outrageous words, but fairly typical of anonymous reviewers on the subject of negative carbon isotope anomalies. The entire review was not significantly longer than the excerpt shown and so of little merit. Peer reviews normally run to several pages or more of detailed analysis. Although peer review has toughened my hide against such barbs, the review process can still be extremely disheartening, especially for students and early-career researchers.

If the negative anomalies truly represent primary isotopic compositions, then their low values mean that a large portion of the carbon dioxide in the surface environment must originally have been organic, which implies net oxidation of organic carbon for millions of years worldwide. When you run the numbers, at least twice as much surplus oxygen would be required as could be produced from organic burial alone. In other words, we need an additional source of oxygen. The oxygen deficit is monstrous and would be impossible today, so where on earth would the extra oxygen come from? Moreover, what was the source of the organic carbon? Or is the opinionated reviewer right that the whole idea is nonsense, and we can simply throw our carbon isotope record out with the garbage?

We dealt in chapter 6 with the so-called Rothman model that envisaged a Proterozoic ocean with unusually high amounts of unreacted organic carbon, which the authors referred to as a dissolved organic carbon, or DOC, reservoir. It seems highly plausible that oxidation of organic matter is incomplete or kinetically controlled even today, so at significantly lower levels of oxygen, it is not hard to conceive that a vast DOC pool could build up, especially in the deep ocean, which over time could wax and wane, causing positive and negative carbon isotope anomalies. Numerical modeling has shown, however, that there isn't enough oxygen, even when you include all of the associated electron acceptors like sulfate, nitrate, or molybdate oxyanions, in today's oceans and atmosphere, let alone during the Ediacaran Period, to oxidize such a reservoir. The problem seemed insurmountable for a long time, but maybe we were looking at the oxygen problem from entirely the wrong angle.

Remember, the past is a different place, so today's earth system is perhaps not where we should be looking for analogies.[8]

In recent years, we have started to turn the problem around. Most earlier models envisaged that massive amounts of organic carbon oxidation can only occur in a modern-like, well-oxygenated earth system, caused, for example, by the actions of newly evolved sponges or other metazoans, like rangeomorphs, which like all animals consumed and oxidized organic matter. Some models proposed the injection of organic matter in the form of hydrocarbon or methane seepage into a modern-like earth system. But what if organic carbon oxidation did not take place in a highly oxygenated environment? What if it was caused instead by a high flux of oxidized species other than oxygen into a largely anoxic ocean? What if the oxidizing capacity of the oceans increased at those times to far greater levels than today, allowing aerobic life forms the opportunity to radiate around the planet? Such an approach has the advantage that oxygen levels per se need not have been high, just the rate of oxidation. This idea also ran into problems, as it implied the abrupt appearance of a continuous stream of surplus oxidant over millions of years without any obvious source. As we saw earlier, oxygen flux to the surface environment today is dominated by carbon burial. However, if carbon burial was high, then carbon isotope values would rise, not fall. In other words, if the oxygen released by excess carbon burial was used to oxidize organic carbon, there would be no net change in terms of oxygen budget or carbon isotopes. However, there is one other major source of oxygen.[9]

A smaller, but significant amount of oxygen is released today by sulfate reduction followed by the burial of iron sulfide, or pyrite. In reality, sulfate reduction to sulfide does not directly produce oxygen. However, the net effect is the same. Sulfate reduction oxidizes organic matter with the help of microbes and so spares free oxygen from being consumed, allowing oxygen released by photosynthesis to remain in the surface environment. Although mole for mole, sulfate reduction releases almost twice as much oxygen as organic carbon burial, pyrite burial is widely believed to be a net oxygen-neutral process on long time scales. This is because oxidative weathering and reductive deposition of sulfide minerals are expected to occur at about the same rates. Much of the sulfate used to oxidize organic matter comes originally from pyrite oxidation during weathering, so it appears that both these mechanisms, organic burial and pyrite burial, are dead ends, borrowing oxygen from Peter to pay Paul, with little or no net oxidizing capacity left over. No wonder there was, and still is, so much skepticism around. But one remaining possibility presents itself, which brings us back full circle to salt.

About half of all the sulfate in rivers today derives not from the weathering of pyrite but instead from the direct dissolution of evaporite minerals like gypsum. Other common sulfate minerals also precipitated from seawater, such as anhydrite (an anhydrous form of calcium sulfate) or polyhalite (a sulfate mineral with potassium calcium and magnesium cations), but for simplicity's sake I will refer to all of these as gypsum as they all behave the same way upon weathering. Unlike pyrite, no oxygen is consumed during gypsum weathering, which releases calcium and sulfate ions in equal measure, although plenty of oxygen can be released following bacterially catalyzed reduction of that sulfate to form pyrite. Today, this is not a major source of oxygen because euxinic conditions are rare and pyrite burial rates are low. In other words, in our modern oxygenated ocean, most sulfate simply stays in the sea or is removed as sulfate. However, this might have been very different at specific times in the past, especially when oceans were anoxic. Gypsum deposits store up oxygen in the form of sulfate

like a battery stores energy and release it only when those sedimentary basins undergo weathering, following tectonic uplift
and erosion of evaporite giant basins. This happened, for instance, when India crashed into Asia, but no extreme oxygenation resulted because the oceans and atmosphere were highly
oxygenated already. As a result, sulfate concentrations simply
increased to their highest ever levels. Then as now, euxinia was
restricted to just a few marginal ocean basins like the Black
Sea. To persuade sulfate to release its oxygen, oceans need to
be both anoxic and productive. This is because microbial sulfate reduction is fueled by organic production. The nutrient-
rich, ferruginous oceans of the Ediacaran Period would have
been the perfect biochemical factory to turn salt into oxygen.[10]

   In the last chapter we looked at the extraordinary tectonic
upheavals of the Ediacaran Period that led to Himalayan-scale
mountains along the Transgondwanan orogenic belt, which
stretched for thousands of kilometers across modern Africa,
Australia, and South America. The Ediacaran-Cambrian transition was clearly a time of enormous sedimentary and, presumably, nutrient input into the world's oceans, fueled by higher
outgassing rates, but what about gypsum? Thanks to studies following Don Canfield's lead, there is plenty of evidence to show
that Proterozoic oceans were anoxic and even widely sulfidic
during the boring billion, whereas prior to the GOE the entire surface environment was devoid of oxygen, and therefore
sulfate. Until recently, you might be forgiven for questioning
whether sulfate ever built up in the oceans before they became
widely oxygenated during the Ediacaran Period. We now know,
however, that some gypsum-bearing evaporite giants were indeed deposited, albeit rarely, before the Cryogenian Period.
Could these have been uplifted and eroded to produce negative carbon isotope anomalies? For the answer to that question
we need to look at much older tectonic events.

The Ediacaran-Cambrian assembly of Gondwanaland merged fragments once part of a supercontinent called Rodinia that came together between about 1,200 and 900 million years ago. For a further hundred million years or more, Rodinia seems to have held together before beginning the long process of breaking apart. In recent years this picture has become more concrete due to improved dating of volcanic eruptions that mark the rifting. It now seems that, despite some earlier failed rifting events, the major, final stages of breakup took place after 750 million years ago and, as discussed in an earlier chapter, were incriminated in the onset of the Sturtian glaciation around 717 million years ago. Just as the Cretaceous evaporite basins were formed during rifting, several other rifting events left major evaporite deposits and the rupture of Rodinia was no exception. Large evaporite deposits of Tonian age (about 820–770 million years ago) are known from Australia, Canada, and Congo that even today exceed a million cubic kilometers, although a much greater volume must have eroded away since then. Surprisingly, these deposits are rich in gypsum, suggesting that oceans, during the late Tonian at least, contained abundant sulfate. A few other, much smaller sulfate-rich deposits are known from even earlier times, 2.0, 1.6, and 1.2 billion years ago, suggesting that Proterozoic oceans experienced far more variability in composition than is generally assumed. A recent detailed analysis of the tectonic settings of these evaporite rift basins concluded that every one of the Tonian gypsum deposits was exposed to weathering during the middle of the Ediacaran Period, precisely when the Shuram anomaly occurred. Could the weathering of these deposits have supplied enough sulfate to oxidize enough organic carbon to have caused the anomaly?[11]

Originally, it had been thought that the anomalies represented the consumption of oxygen. However, as we have seen in a previous chapter, oceans became oxygenated around this time and sulfate concentrations even increased during negative excursions. In other words, we need to turn this idea on its head and assume that oxidizing capacity was being provided by sulfate in excess of what was required to oxidize the organically turbid "Rothman" oceans. Assuming equilibrium, this means that surplus sulfate needed to be provided at approximately half the rate that organic carbon was being oxidized, and for several million years. This is because one molecule of sulfate releases almost two molecules of oxygen and can therefore oxidize almost two atoms of organic carbon to carbon dioxide, following pyrite burial. The results of our mass balance calculations surprised even us when they showed that the sulfate input from rivers only needed to have been two or three times higher than today to produce negative carbon isotope anomalies. The key point here is not the flux itself but that most of that sulfate needed to come from salt deposits and be buried as pyrite.

The jigsaw puzzle pieces of this tale are finally falling into place but have grown quite complex and numerous. They include tectonics, weathering, nutrients, organic carbon, gypsum, pyrite, oxygen, climate, and, of course, life. Although assuming mass balance can quantify the sulfate needed to generate the anomalies, numerical modeling is the key to following the entire process. The first simple biogeochemical models were published in the same year that the silicate weathering feedback was being proposed by a trio of American atmospheric scientists. Two years later, in 1983, a different American trio proposed the computer model BLAG, named for the initials of their names: Berner, Lasaga, and Garrels. Their approach quantified long-term carbon cycling by using rate equations for each link in the cycle and assuming steady-state conditions. BLAG evolved into GEOCARB, which is still widely used today and has the main difference that carbon isotopes, and later sulfur isotopes, are used to drive the model outputs, which are generally atmospheric oxygen or carbon dioxide levels. Many an author has used the resulting curves to bolster one or another theory of radiation and extinction. GEOCARB has been immensely beneficial to the geological community and continues to be developed, but with significant caveats.[12]

"All models are wrong, but some are useful," said some oft-cited wise spark, and GEOCARB is no exception. Because it takes a reverse modeling approach, the outputs are driven by the data and so are dependent not only on the quality and resolution of the data, but on our understanding of what they mean. As we have seen, carbon isotope anomalies cannot be understood using conventional mass balance, which means such data-driven models result in implausible or even impossible results, such as negative oxygen levels. For this reason, forward models were developed, foremost among them being COPSE, the brainchild of Noam Bergman, Tim Lenton, and their supervisor Andy Watson at the University of East Anglia. COPSE is different from GEOCARB in that the model predicts the isotope data, and so uses them to test rather than drive the model outputs. Examples of inputs into the model are orogenies, continental configurations, evolutionary events, and other phenomena inferred from the rock record. The parameterization of the models is often the same with both types of model, but the advantage of forward models is that interpretations can be verified using a known dataset, in this case the isotope record. For our work on the Shuram anomaly, Ben Mills, another student of Andy Watson who is now an associate professor at Leeds University, and I used the COPSE model forward approach to test the possibility that gypsum weathering could produce negative carbon isotope anomalies, while at the same time not consuming, and potentially adding to, the amount of free oxygen. The main thing we needed to do was remove the inherent but unrealistic assumption in all such models that gypsum weathering rates are precisely equal to gypsum deposition rates on all time scales, thus freeing up oxidizing capacity.[13]

Tectonic events and the aftermath of the Snowball Earth seem to have fueled high rates of weathering. This is because collisions caused uplift, while metamorphic and volcanic carbon dioxide weathered the exposed rocks chemically. The released nutrients fertilized the oceans, turning highly productive, euxinic ocean margins into pyrite factories as long as they were supplied with sufficient sulfate, which they were, courtesy of the uplift and erosion of specific basins with huge gypsum deposits. Numerical modeling shows that a relatively modest input of gypsum can generate the requisite oxidizing capacity, and so the mystery of the Shuram anomaly and perhaps other such anomalies dissolves away. There is simply no need to break the back of the earth's oxygen budget if ancient sulfate deposits get turned efficiently into sulfide minerals. Counterintuitively, it even helps for the world to be less oxygenated in order to drive the requisite surplus oxidation. The result, oxygenation of the oceans, caused one of the most astonishing biological radiations of all earth history when sessile, benthic, fractal rangeomorphs, the poster fossils of the Ediacaran Period, exploited the newly oxygenated seafloor down to great depths. Ben's further modeling adds nuance to the picture, in the form of predictable knock-on effects.

The COPSE modeling done by Ben suggests that none of this would work as envisaged if it were not for the generation of a key positive feedback. The reason why this is needed becomes clear when we think of the effects of an oxygenating ocean on the stability of aqueous sulfide. The reason we would not experience such a cascade of events today is that oxygenation would shrink the extent of euxinia and so stop pyrite burial and oxygenation in its tracks. To understand why this did not happen during the Ediacaran, we need to think again about the vast reservoir of organic carbon in the Neoproterozoic ocean. The Shuram anomaly informs us that sulfate reduction, likely catalyzed by microbes, was coupled with carbon oxidation, catalyzed by animals. This carbon oxidation would help to raise carbon dioxide levels in the atmosphere, stabilizing warm climates and high rates of chemical weathering. It is this feedback that would have sustained sufficient riverine input of sulfate and nutrients to drive pyrite burial, thereby prolonging oxygenation for as long as there was organic carbon to oxidize and sulfate to weather (Figure 24). Even with this feedback operating, it is evident that not all sulfate was reduced to sulfide, as seawater sulfate and other oxyanion concentrations even increased throughout this period. It is telling that the sulfur isotope composition of the oceans fell dramatically at this time to match precisely the values of those Tonian evaporite deposits.[14]

The curious case of the Shuram anomaly suggests that the DOC reservoir acted as a huge capacitor buffering against perturbations to the earth's atmospheric composition. Unlike today, when it is the silicate weathering feedback and the inorganic carbon cycle that regulate climate, the organic carbon cycle was likely the more important climate regulator in the distant past. As long as a large fraction of surface organic carbon remained reduced, it could act as a vast oxygen capacitor, soaking up excess oxidant. During cooling episodes, the greater oxygen solubility of cold water would have altered the balance between old and new carbon in favor of the new, releasing carbon dioxide in the process, and keeping the planet warm. Perhaps this is the reason why the boring billion years of equable climate and muted oxygen were so stable, and carbon isotope deviations from the norm were so rare and transient. The fact

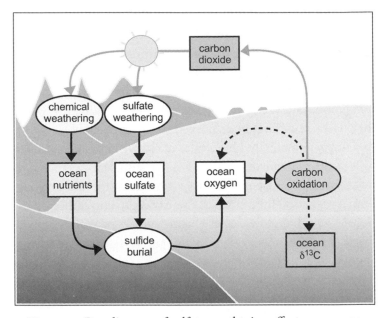

Figure 24. Box diagram of sulfate weathering effects on oxygen and climate, showing the natural feedbacks in the carbon and sulfur cycles when evaporite sulfate minerals such as gypsum dissolve following tectonic uplift and erosion during the Ediacaran Period. Boxes represent reservoirs, ovals represent processes; solid arrows show positive effects, and dashed arrows show negative effects. Warmer climate, caused by the carbon oxidation and subsequent greenhouse effect, leads to greater chemical weathering, nutrient input, and organic production, and thus a positive feedback toward increased pyrite (sulfide) burial, following bacterial sulfate reduction, and ocean oxygenation.

that they did occur suggests the existence of several *failed* oxidation events before the Cryogenian Period.

Having been few and far between, negative carbon isotope excursions started to come thick and fast once Rodinia began properly to break apart at about 750 million years ago, becoming both frequent and extreme after about 650 million years ago once rifted cratons began crashing into each other. One might think that the organic reservoir would have prevented such carbon cycle perturbations from causing climate change, and this seems valid, but the sheer magnitudes of some of the negative anomalies suggest that the organic capacitor must have neared exhaustion. Each Snowball event was preceded by anomalously deep negative excursions, and the extreme Shuram anomaly also led into a regional glaciation. This is precisely what we would expect were the Proterozoic Earth System buffered by the ocean's store of organic carbon, while the gradual disappearance of such negative anomalies by about 500 million years ago—that is, the late Cambrian—suggests further that the reservoir of reactive organic carbon gradually waned away as oceans cleaned up their act through the ventilating actions of the newly evolved animals, which either had guts to turn organic carbon into dense fecal packages or diverse apparatus to filter it out. The story of the Cambrian explosion is about much more than oxygen, however. Can evaporite dynamics tell us about other aspects, such as the high rates of evolutionary change, the emergence of energetic metabolisms, or the onset of biomineralization? Indeed, why did animals build shells in the first place? And why use calcium phosphate and then calcium carbonate as the first biominerals?

An early explanation for why the first biominerals were calcium compounds recalls that calcium is toxic to animals. Perhaps, it was postulated, animals evolved in marine environments in which calcium concentrations were increasing steadily to dangerous, even lethal levels. Excreting calcium in the form of a shell was therefore a healthy thing to do. Once shells came about, claws and teeth needed to keep up, escalating the arms race. Other creatures then also needed to produce minerals for protection, and sponges chose silica, which is the other major biomineral today. One of the quirks of biomineralization is that most skeletal animal groups retain their original mineralogy even when it would consume less energy to make shells from a different mineral. The first calcium biominerals appeared during the recovery phase of the Shuram anomaly and may have been only lightly rather than actively controlled by their biological hosts. Therefore, we might suppose that the onset of biomineralization by calcium minerals was preceded by very high calcium levels in seawater. Does this correspond with the increased gypsum weathering model?[15]

Weathering brings not only nutrients but also calcium ions into the oceans, and it has long been speculated that high rates of weathering, therefore, might explain the onset of biomineralization. However, this argument has always had a potentially fatal flaw. Although calcium ions are predominantly delivered into the ocean by rivers, chemical weathering also brings in carbonate ions. Because the oceans are always saturated with respect to calcium carbonate minerals, the extra calcium would simply combine with the extra carbonate, and precipitate out, leaving calcium levels in the oceans close to where they were previously. This is a bit like Justus von Liebig's barrel metaphor. Liebig's famous barrel had staves of different heights, so its capacity was dictated by the height of the shortest one, no matter how much beer was poured in. Unless carbonate deposition was somehow restricted by other factors, like having enough "accommodation" space, there seems to be no reason why higher fluxes alone would lead to higher concentrations of calcium, and certainly not to toxic levels. However, this logic does not hold in the case of sulfate mineral weathering.

When gypsum dissolves during weathering, no carbonate alkalinity is consumed or released. This means that the weathering flux is imbalanced in favor of calcium, and so proportionally less calcium is deposited than carbonate, resulting in a higher overall calcium concentration in the ocean. There are two additional reasons why calcium concentrations would have risen around this time. First, following the Shuram anomaly, sea levels seem to have fallen for a time, possibly due to the cooling effects following exhaustion of the organic carbon reservoir. An exquisitely preserved glacial pavement and overlying tillite at Luoquan in North China marks the regional glaciation around this time. Growth of ice sheets and cooler oceans would have resulted in less accommodation space on the continental shelves where carbonate minerals could be deposited, and a concomitant rise in the calcium carbonate saturation state of the ocean. Second, and more generally, oxygenation of the seafloor would have caused dissolution of freshly deposited carbonate, raising the activities of both calcium and carbonate ions in bottom waters. It would appear to be a reasonable assumption, therefore, that the latest Ediacaran was a time of unusually high calcium concentrations, especially close to the interface between sediments and overlying seawater. All three mechanisms are related to the evaporite dissolution–pyrite burial model of ocean oxygenation, so the onset of biomineralization and even our own backbones may be the result of evaporite dissolution.

Previously, it had been thought that ocean oxygenation was a one-way street and that once Nursall's threshold had been overcome it was all plain sailing thereafter. However, as I noted earlier, nothing could be further from the truth. Uranium isotopic work has shown that anoxia was exceptionally widespread once again in the latest Ediacaran, and that bilaterians became increasingly restricted to marginal oxic environments before the start of another series of biotic radiations (and crises) during the explosive Cambrian radiations. With so much oxygen released by evaporite weathering, how was it that things went into reverse so soon? There are two main explanations for this reversal that hold water.[16]

First, we need to think back to the positive feedback that drove oxygenation in the first place. It was not only the gypsum weathering that achieved this, but importantly the release of carbon dioxide, which helped to push productive ocean basins euxinic by enhancing global climate, chemical weathering, and nutrient input. However, the recovery of carbon isotopes back to seeming normality suggests that some link in this chain of events was no longer in place. It could be that the ocean organic reservoir had become too small. It could also be that gypsum weathering rates had become lower than deposition rates, possibly for the first time since the previous evaporite giant event 200 million years earlier. Both of these could be true, but we have direct evidence for the second. The start of widespread evaporite deposition is marked at 550 million years ago, and it continued through to the Cambrian Period. Massive gypsum deposits of this age have been found in China, India, Pakistan, Oman, and Siberia. Estimates of their original volumes come in at twice the size of those that formed during the Miocene salinity crises. Just as in the Cenozoic, massive weathering eventually led, once again, to massive deposition.

Not only would this have stopped oxygenation, it would potentially have caused deoxygenation because some of the oxygen consumed during oxidative weathering could not be returned by pyrite burial, causing an oxygen deficit on long time scales. Remember that all sedimentary pyrite exposed to weathering since the GOE has been entirely oxidized by the earth's oxygenated atmosphere. Surplus burial of gypsum leads, therefore, to a steady leak of oxidizing capacity from the earth system. In the more reducing environments of the deep past, this leak would have had a much greater effect, especially once sulfate concentrations became limiting to pyrite burial. Our sulfur isotopic work has shown that sulfate concentrations declined steadily through the early Cambrian, reaching minimum values by the time of the first mass extinction of the Phanerozoic, at around 515 million years ago.

The existence of animals during the Cambrian Period ensured sharper redox boundaries in the oceans, permitting the existence of animals at all times in shallower, more oxygenated environments. When conditions improved, they radiated throughout the oceans, occupying the same wide range of ecological niches that animals enjoy today. But when conditions deteriorated, competition must have been fierce as both Ediacaran and Cambrian forms of animals clashed in the few available, survivable realms left. This environmental instability likely fueled the burst of innovations we recognize as the Cambrian explosion. Each crisis likely caused an evolutionary bottleneck through which only a depauperate community was allowed to pass. The result was an exceptional period of diversification.[17]

In recent years, the recognition that the Ediacaran-Cambrian transition was marked, not by a single boom, but by a series of booms and busts, has provided a convincing but hitherto unsuspected resolution to Darwin's thorny dilemma over the Cambrian explosion. Although tectonic upheavals provided the backdrop to these events, their role was largely unforeseen even a few years ago. Yes, tectonic uplift brings in nutrients and detritus that can help to produce and bury organic matter, generating oxygen, but it also leads to greater exposure of reduced organic carbon and mineral sulfide to oxidation. Their oxidative weathering consumes surplus oxygen, so it is wrong to interpret massive perturbations to climate, nutrients, oxygen, and the global carbon cycle, viewed through the lens of unprecedented carbon isotope instability, simply as a result of tectonic uplift or erosion. Instead, it is the role that tectonics played in the global sulfur cycle that may have been equally, if not more, important. Imbalances between the weathering and deposition of sulfate and sulfide minerals can defeat the negative feedbacks that are thought to govern the earth system on long time scales. Those feedbacks we have teased apart, like the DOC capacitor or the silicate weathering thermostat, are believed to soak up imbalances in the carbon cycle on all time scales, but we now know that some of them can be overpowered by sustained imbalances in the much smaller sulfur cycle. Such imbalances may explain how Snowball Earth occurred, and in turn may explain the circumstances that led not only to the diversification, radiation, extinction, biomineralization, and locomotion of all animals but even to their evolutionary origins. I don't think there is any way I could have guessed how our understanding of these events would change so quickly when I embarked on my adventures in China, Mongolia, and Mauritania more than twenty years ago, but it has been a wild ride to get this far.

Although we appear to have come full circle in our story of the evolutionary origins of animals, there is still a bit more to tell. In the final part of the book, we will explore whether sulfur cycle imbalance could drive climate change, oxygenation and biological radiations, diversification, and extinction at other times in earth history. Could it be that evaporites have yet more secrets to reveal?

11

Biosphere Resilience

Solar and tectonic drivers forced changes to the earth system, allowing those forms of life most suited to the new ecological opportunities to spread around the world. However, animal life took many hits during subsequent mass extinction events when climate change and ocean anoxia combined to reset biological evolution. Sulfur cycle forcing and feedbacks likely played a role in such catastrophes, and in shaping our world ever since the Great Oxidation Event 2,300 million years ago.

After years of globe-trotting, it was to University College London on Gower Street that I finally returned. In the intervening two decades, I had been incredibly lucky to live and work in Zurich, Strasbourg, Ottawa, Townsville, Münster, and Nanjing. My fieldwork had taken me from Scotland to China, Mongolia, Mauritania, Mali, Senegal, Canada, the United States, India, and the Australian outback, before returning full circle back to China and Scotland. Walking past the biology department one day, I discovered, courtesy of a blue plaque near its entrance, that Charles Darwin, on returning from his own travels, had also made Gower Street his home. It was there that he wrote down most of his geological ideas, such as his new theory for how island atolls had formed around sinking volcanoes. It is also where he cautiously began to write down his then controversial ideas about our evolutionary origins. But Gower Street's hustle and bustle proved unbearable, and he soon left Macaw cottage, jokingly named for its garish interior, for Down House in the Kent countryside. Most years I take new members of my research group on a trip to Down House. It is well worth a visit just to walk the "thinking path" he strolled along each morning and afternoon. Sauntering through the home and grounds of a typical nineteenth-century country squire, it is not hard to sense the restful inspiration that helped him write. After nearly completing this book, and at the same age he was when he completed the *Origin of Species,* I am filled with awe, and not a little envy, at how he was able to write so much, so coherently.[1]

After sailing around the world, Darwin returned to a city that was rapidly changing. Apart from all the building construction Darwin hated, London was becoming a focal point of radical geological thinking. It centered around the Geological Society, which had been meeting at Covent Garden's Freemasons' Tavern for almost thirty years. Rather more notably, the tavern also hosted the meetings that established the rules of football. If only we could journey back in time for a quick pint, we might be regaled at the bar with voyages of discovery and radical evolutionary theories on one night, only to hear heated exchanges about the offside rule on the next. By the time Darwin moved away from dusty Gower Street in September 1842, he had already presented four papers to the society about his voyage on the *Beagle,* been elected to its council, and been appointed its secretary. George Bellas Greenough, author of the second geological map of Britain, was an active founding member of both the society and UCL. It was Greenough who saw to it that geology was taught at Gower Street since the university's beginnings in 1826. In those days, UCL was simply called London University, having been conceived as a secular alternative to the establishment bastions of Oxford and Cambridge. By 1831, the year Darwin set out, John Phillips, nephew and mentee to William Smith, author of the first geological map of Britain, had begun to give lectures at the new London University. Smith and Greenough had not only made rival maps but also had conflicting ideas about whether fossils could be used for correlating sedimentary layers, or strata. The outcome of that debate, in favor of Smith's new science of *biostratigraphy*, would greatly influence Darwin's thinking. However, by the time Darwin returned five years later, London University had become a mere college of the new University of London, a subtle but significant change, while Phillips had moved on.

The reason for the change was that while Darwin was away, Oxford and Cambridge petitioned the king and Parliament to refuse London University a royal charter to award its degrees. King George IV then set up his own King's College, leading John Phillips and others to leave the "Godless Institution" in Bloomsbury for kinglier, and presumably godlier, things. In 1836, the depauperate upstart had to beg the new king's condescension to gain a royal charter and so became a sibling college of a new, overarching University of London. I don't know what Darwin would have made of these political machinations in his absence, other than perhaps to rue the building work that resumed in earnest after his return, but I relate this short diversion because it is an example of how today's world depends on past events. In vainer moments, I like to imagine that, if the king had been a little more magnanimous, John Phillips might have been my own predecessor as the first professor of geology on Gower Street, rather than a mile down the road on the Strand, when he published the very first geological time scale in 1841. The flight of time's arrow is contingent on the diverting events of history, and as Darwin was allowing himself to think, earth history is no exception. Just three months before moving away from Gower Street, he completed a thirty-five-page "pencil sketch," the first written version of his theory of evolution by natural selection.

Phillips based his time scale not on absolute time, but on relative time, using the sequence of fossils he had painstakingly reconstructed over many years. He was helped in the task by a long geological apprenticeship with his guardian William Smith. In his youth, Phillips had accompanied Smith on his travels around Britain and was understandably more influenced by the ideas in Smith's pioneering book *Strata Identified by Organized Fossils* than by establishment skeptics like Greenough. William Smith taught how diagnostic fossil assemblages could be used to trace strata from one area to the next, and it was this new biostratigraphy that helped Phillips divide fossiliferous strata into three distinct ages, which he called the Palæozoic, Mesozoic, and Cænozoic Eras, each one separated by an evident decrease in the number of fossil taxa. Smith's biostratigraphy also helped Darwin trace our evolutionary origins back to the invertebrate animals of the Cambrian Period, which still marks the start of the Palæozoic Era. Neither Phillips nor Darwin speculated much about Precambrian strata, which their contemporary Adam Sedgwick called the Protozoic. In North America, where older, Precambrian rocks are more common, geologists had identified two distinct series of barren, or *Azoic*, rocks, the Huronian and the Laurentian, which later became the Archean and Proterozoic Eons, respectively. Although Darwin and many of his contemporaries favored gradual over catastrophic change, believing abrupt changes to be the result of an incomplete rock record, we now know there to have been both gradual trends and genuine extinction and radiation events throughout earth history. Today, we know Phillips's three major Phanerozoic transitions to be the Cambrian radiations and the Permian-Triassic and Cretaceous-Paleogene mass extinctions. Such evolutionary radiations and bottlenecks changed the course of biological history and are the subject of this penultimate chapter.[2]

As well as the two era boundaries recognized by John Phillips, later geologists have identified a further four major mass extinctions, at the end of the early Cambrian radiations and near the end of the Ordovician, Devonian, and Triassic Periods. The Cretaceous-Paleogene boundary is marked by a meteorite impact, which left a crater in Mexico 150 kilometers wide and 20 kilometers deep. However, despite years of searching, no other mass extinction can be linked to an extraterrestrial cause. The other five mass extinctions are all associated with climate change and ocean anoxia. Four of them coincide with volcanic eruptions, so greenhouse forcing has been proposed as the most likely underlying cause. Some if not all of these extinctions also follow times of major evaporite deposition. Over the past few years, I have pondered whether evaporite giants might hold clues, not only to the evolution and diversification of animals during the Cryogenian and Ediacaran Periods, but also to the numerous extinctions to which aerobic life was subject during later times. This is because imbalances between evaporite deposition and weathering have the potential to reinforce or destabilize the carbon cycle feedbacks that are thought to govern both oxygen and climate. Mass extinctions, like Snowball glaciations, were times when regulatory mechanisms were seemingly weak or absent. Without the evaporite piece of the puzzle, we may be misinterpreting some of the most influential events in earth history. To understand why, we need to take a closer look at the Miocene Epoch and the salinization of the Mediterranean Sea.[3]

In the previous chapter, I emphasized just how big a deal the Messinian Salinity Crisis was, leading as it did to a million cubic kilometers of salt being deposited in just half a million years. Amazingly, however, the MSC may not even have been the biggest gypsum deposition event of the Miocene Epoch, as it followed hot on the heels of a similarly huge event caused by seawater spilling into a deep tectonic fissure that was destined to become the Red Sea. The biggest Miocene event, however, was probably even earlier, and related to the closure of the Mesopotamian Seaway, which separated the central Tethys (Paratethys) from the eastern Tethys Ocean. Evaporite deposition began in earnest around fifteen million years ago and led to hundreds of meters of gypsum throughout Iran and the Middle East. Nowadays, oil companies target these impermeable salt deposits because they form ideal traps for rising petroleum. The closure of the Mesopotamian Seaway culminated in the Badenian Salinity Crisis that began around 13.8 million years ago across eastern Europe. Although there had been few, if any, evaporite giants since the opening of the Atlantic Ocean more than one hundred million years earlier, during a short, ten-million-year interval of the late Miocene Epoch, there were no fewer than four massive events, one after the other. In fact, so much salt was deposited during the late Miocene that ocean salinity decreased globally by more than 10%, with even greater changes to specific elements in seawater. This is because of the predictable precipitation sequence from calcium carbonate, to calcium sulfate, and finally to sodium chloride and other minerals. As a result, Ca concentrations would have been most affected, declining by far more than 10%, and possibly by as much as 50%. Curiously, until now few people have taken such extreme but predictable changes into account.[4]

The Miocene Epoch is still as much debated today as when I was studying for my PhD a quarter century ago. It is not without trepidation, therefore, that I wade into such hot waters, as I am liable to be challenged by Cenozoic experts who know far more about that period of earth history than a Precambrian geologist such as myself. However, it would not be the first time. One of the earliest research projects I was involved in, in Zurich, focused on Miocene phosphatic limestones of Malta. It arose out of some strontium isotope measurements I performed for a fellow student, and I became so interested that I collected more material when I visited Malta myself in 1995. The project concerned the carbon and oxygen isotope compositions of benthic and planktic foraminifera, which form tiny whorl-like shells composed of calcite. My main role in the project was to use the strontium isotope composition of the shells and other material to date the stable isotope features by comparing them against the global seawater curve. By comparison with those who have spent their entire careers reconstructing climate change using the same oxygen isotope paleothermometer, and at much higher resolution, I was merely dipping my toes in. Nevertheless, the study went well enough that we were able to identify and date a well-known shift during the middle Miocene to much cooler conditions, as indicated by an abrupt increase in the $^{18}O/^{16}O$ ratio of the Maltese foraminifera.[5]

The *mid-Miocene climate disruption* is still poorly understood but thought to have led to a wave of extinctions of both terrestrial and aquatic life forms, including crocodilians, between about 14 and 13.5 million years ago. What we do know, however, is that it was associated with the onset of a period of steady cooling that led to a huge expansion of the East Antarctica Ice Sheet, causing sea levels to fall precipitously. The mid-Miocene disruption followed fifteen million years of relatively stable, warmer conditions, and was in turn followed by episodic cooling until the present day. Although ocean circulation changes and organic burial have both been suggested as potential triggers for the cooling, it is fair to say that there is no firm consensus about why our planet suddenly cooled around that time. Boron isotope studies implicate a fall in atmospheric carbon dioxide, but we don't know what caused that either. It is intriguing, though, that the mid-Miocene disruption coincided perfectly with deposition of the largest gypsum deposits for more than 100 million years during the closure of the Mesopotamian Seaway. Moreover, further evaporite deposition, this time in the Red Sea and Mediterranean Sea, accompanied the other major cooling episode of the Miocene, during the Tortonian-Messinian interval, which led to expansion of ice sheets on Antarctica to their modern levels and the onset of northern hemisphere glaciation. The mid-late Miocene interval was indisputably the climatic turning point that nudged our planet toward deeper glaciation.[6]

One possible link between evaporite deposition and climate is that, in a glacial world in which ice sheets can wax and wane, global cooling might trigger evaporite deposition by causing ice sheets to expand, promoting seaway restriction amid falling sea levels. However, in the case of both the Miocene evaporites, deposition began before the major ice sheet expansion, although it was certainly exacerbated by later sea level falls. Could it therefore work the other way around as well, with evaporite deposition causing climate change? We saw in the previous chapter that sulfate weathering could lead to global warming by contributing to the oxidation of organic matter in an anoxic ocean, and the reverse situation might also work, whereby sulfate deposition could lead to a buildup of organic carbon and cooling. Although this does not seem a likely mechanism in a fully oxygenated world like the Miocene Epoch, there could be another way for evaporites to disrupt the long-term carbon cycle, via the calcium cycle.

You may recall that the earth's weathering thermostat helps to regulate climate via the temperature dependence of silicate weathering. At higher temperatures, rates of silicate weathering increase, releasing more calcium ions and carbonate alkalinity into the ocean where they will eventually recombine to form calcium carbonate. Because silicate weathering consumes carbon dioxide, calcium carbonate precipitation acts as a net sink for carbon dioxide, thus pushing back against the initial rise in global temperatures. For every mole of calcium released to the oceans, one mole of carbon dioxide is stored for a geological cycle or more. That is the theory, at least. But is this balancing act always as perfect as exponents of the long-term carbon cycle have presumed since its discovery in the early nineteenth century by the great ceramicist and polymath Jacques-Joseph Ebelmen? It is healthy to question all such fundamental premises, and it is great fun to do so even if they survive the attention. In recent years, a number of geochemists have begun to chip away at the silicate weathering paradigm by looking at the two other major sources of calcium to the ocean that are not linked to removal of carbon dioxide.

The major source of the calcium in seawater is not in fact silicate weathering but limestone weathering, by some margin. However, on time scales longer than about a few thousand years carbonate weathering and deposition ought to simply balance each other out. In other words, limestone weathering is neither a net source nor a net sink of carbon dioxide, and so ought not to greatly influence climate. There are exceptions to this rule, most notably if carbonates are not weathered by rainwater, which is a dilute form of carbonic acid, but instead by sulfuric acid released during the weathering of pyrite. In such cases,

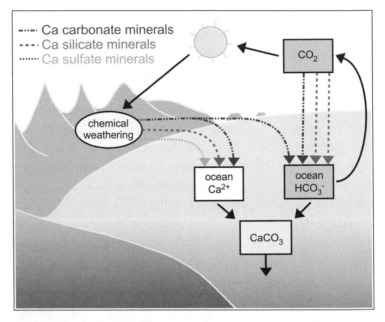

Figure 25. Box diagram of calcium weathering effects
on climate, showing how chemical weathering of calcium-
bearing minerals provides the constituents to make up calcium
carbonate, which ends up deposited on the seafloor. Each pathway
has different implications for atmospheric carbon dioxide, and
therefore for climate. Following the arrows shows that silicate
weathering leads to a net loss of carbon dioxide, carbonate
weathering results in no net change, and sulfate weathering
causes a surplus of calcium ions over carbonate alkalinity.

limestone weathering can even become a net source of carbon
dioxide with relatively minor effects on climate (Figure 25).

The third source of calcium, after limestone and silicate
weathering, is gypsum weathering, which as we have seen is
not balanced by gypsum deposition even over millions of years.
As a consequence, calcium ions from dissolved gypsum enter
the ocean unaccompanied by carbonate alkalinity, but usually

leave it as calcium carbonate, rather than gypsum, thereby un-balancing the oceans' alkalinity budget. The net result of gyp-sum weathering, therefore, is to alter the ratio between calcium and carbonate ions in seawater in favor of calcium. With fewer carbonate ions in seawater, aquatic carbonate species re-equil-ibrate in favor of carbon dioxide, some of which degasses into the atmosphere. Intriguingly, this means that when erosion rates are high, the sulfur cycle, in the form of both pyrite and gypsum weathering, acts to counter the cooling effect of in-creased silicate weathering during mountain building, such as would have been caused by the rise of the Himalayas. Although of academic interest among the weathering community, such climatic effects are small beans relative to the potential effects of evaporite deposition.[7]

Today, about half of the sulfate making its way to the ocean derives from pyrite weathering, and the other half is the product of gypsum dissolution. Although some pyrite must be leaving the ocean, it is hard to quantify how much, while there has essentially been no net gypsum deposition since the end of the Messinian Salinity Crisis more than five million years ago. Between about 120 and 15 million years ago, there were also no major evaporite deposition events. This means not only that the sedimentary sulfur cycle is tremendously out of bal-ance, but also that the relatively short deposition events were on a far grander scale, in terms of both flux and magnitude, than weathering events. In a reversal of the warming effects of gypsum weathering, gypsum deposition removes calcium that would otherwise have been deposited as calcium carbonate. This alters the calcium versus carbonate balance in favor of carbonate, increasing ocean pH and lowering carbon dioxide in the atmosphere. The cooling effect of deposition is more abrupt, more severe, and longer lasting than the warming ef-fect of weathering. Ben Mills of Leeds University and I have estimated that major deposition events, if unchecked, could cool the planet by several degrees Celsius. In a glaciated world, evaporite deposition might also entice global climate into a positive feedback, whereby cooling leads to ice sheet expansion, greater basin restriction due to sea level fall, continued evaporite deposition, and further cooling. The astonishing upshot is that our modern glaciated world may be in part contingent on the closure of the Tethys Ocean and related salinity crises.

Assuming you accept my argument that gypsum deposition causes cooling, what, if anything, does this have to say about mass extinctions? After all, although evaporite deposition both preceded and accompanied mass extinction events, most are not believed to be associated with cooling, but instead with ocean anoxia and global warming. The answer to that question relates to *biosphere resilience*, and why at specific moments in earth history, positive feedbacks may race unchecked, just as happened in earlier times during the run-up to Snowball Earth. To understand what this link might be, let's take a closer look at the greatest of all mass extinctions, at the Permian-Triassic boundary, which sounded the knell for the once hugely successful trilobites and many other marine invertebrate species. It is estimated that well over 90% of all marine species went extinct across the boundary, while the toll for terrestrial vertebrates, at about 70%, is only slightly less shocking. Both terrestrial and marine ecosystems were profoundly perturbed. There was a five-million-year gap in gymnosperm forests and coal deposits, possibly caused by excess UV rays, as tree pollens from this time show remarkable mutations thought to be caused by the weakening of the ozone layer by volcanic sulfurous emissions. In the sea, the reef gap seems to have lasted even longer, as the types of organisms that created late Permian reefs, like rugose (finger) and tabulate corals, fusulinid foraminifera, and many other types of skeletal invertebrates, disappeared forever. No wonder that John Phillips chose this time as one of only two era boundaries.[8]

The Permian-Triassic boundary event was the greatest of all mass extinctions and is suitably believed to have been caused by the eruption of the world's largest igneous province, the Siberian Traps. The eruptions likely expelled huge amounts of carbon dioxide into the atmosphere, warming the planet, lowering oxygen solubility, and turning swaths of the oceans anoxic and euxinic. One of the leading experts on the extinction, Michael Benton of the University of Bristol, has tried to address the thorny question of how mass extinctions affected life both on land and in the sea. His solution lies in the extreme nature of the warming, which during the Permian-Triassic interval reached lethal levels. In the sea, a combination of heat and anoxia reduced the habitable depth zone for plankton to an absolute minimum, while on land temperatures as high as 35–40 degrees Celsius caused most plants and animals to suffer major physiological damage. Although greenhouse forcing explains these hostile conditions to some extent, the runaway nature of warming across the boundary suggests that regulatory feedbacks were too weak to normalize climate. Wouldn't we expect increased silicate weathering to pull runaway temperatures back from the breach? And wouldn't nutrient feedbacks sustain organic burial and so prevent anoxia from becoming life-threatening?[9]

For two reasons, the Permian-Triassic crisis came at a challenging time for climate regulation. First, most of the world's cratons were stitched together to form a single supercontinent, Pangaea, which had by then been denuded down to a flat and arid landmass. Low denudation rates meant that weathering was limited not so much by temperature but by the availability of material to weather, crippling the silicate weathering feedback. This is termed transport-limited (or supply-limited), as

opposed to kinetic-limited chemical weathering, and seems to
have been a feature of all long-lived supercontinents. The consequence of supply-limited weathering is that volcanically in-
duced global warming can continue relatively unchecked by
increased chemical weathering rates. Just such a faulty thermostat might also have been a factor in runaway warming after
Cryogenian glaciations and may have contributed to the elasticity of carbon isotope perturbations during the rifting phase
of that other very long-lived supercontinent, Rodinia. Intriguingly, the other period of wild isotopic excursions appears to
have been the post-extinction early Triassic world.[10]

Second, although the tendency for evaporites to form in
hot and arid climates could potentially provide a complementary negative feedback, any cooling effect would be negligible if
sulfate concentrations were too low for gypsum to precipitate.
In Europe, the Permian-Triassic boundary is marked by the
transition from the Zechstein salt to the overlying Buntsand-
stein (colorful sandstone). The impermeable Zechstein salt layers trap much of the oil that is currently drilled in the British
and Norwegian North Sea and has been touted as a possible
repository for nuclear waste. It consists of thick deposits of
halite, anhydrite, and carbonate, stretching across more than
a million square kilometers of northern Europe through Germany into the British Isles. Although the amount of sulfate
in seawater was at almost modern levels during much of the
Permian, leading to huge gypsum deposits around the world,
by the end of the period concentrations had crashed to very low
levels, precluding any possibility of significant gypsum deposition. The absence of both of these negative feedbacks during
a period of almost uniquely massive volcanic eruptions across
Siberia likely played a key role in letting temperatures spiral
during the lethal early Triassic hothouse.

Exceptionally low sulfate, warming, anoxia, and extinction have been a recurrent theme ever since the early animal radiations of the Ediacaran-Cambrian transition. While a student of mine at UCL, Tianchen He reported sulfur isotope results from Siberia showing that seawater sulfate levels decreased from their high point after the Shuram anomaly through the Cambrian radiations, to reach exceedingly low concentrations around the time of the early Cambrian Sinsk (or Botoman-Toyonian) extinction event, which was the first of the six major mass extinctions of the Phanerozoic Eon. Low sulfate concentration is also a common feature of the late Devonian and end-Cretaceous mass extinctions, as well as the end-Permian biotic crisis. Tianchen, now at Leeds University, has more recently demonstrated that this also holds true for the end-Triassic extinction as well, completing the set of warming-related mass extinctions. Only the end-Ordovician event might have been different, but that event is generally linked to glaciation rather than global warming. Most of these biotic events were preceded by massive deposition of gypsum and accompanied by the expansion of anoxic, sulfidic waters, both of which drew down sulfate levels enough to preclude much evaporite-related cooling during the extinction interval itself. The climate effects of evaporites are, however, not the only connection to extinction. In order to tease apart further connections between the sulfur cycle and extinction, we need to look again at how the sulfur and oxygen cycles interact.[11]

As we saw in the previous chapter, the large amounts of sulfate in the ocean today derive from two main sources, pyrite and gypsum weathering, as well as, to a lesser extent, volcanism. Crucially, both volcanic outgassing and pyrite weathering deplete the atmosphere of oxygen, which needs to be returned following pyrite burial. However, because gypsum deposition events deplete the ocean of sulfate, they limit the amount of pyrite that can be buried, causing a net loss of oxygen to the system. Some of this might be made up by extra organic burial as the ocean floor turns progressively more anoxic, thus raising carbon isotope values, but making up the difference would become increasingly difficult. More widespread anoxia also favors euxinia, removing even more sulfate via pyrite burial, and eventually the low sulfate concentrations limit both gypsum and pyrite burial rates. Almost all mass extinctions were preceded by this steady leak of oxidizing capacity from the surface environment, preconditioning the oceans to turning lethally hot and anoxic following massive volcanic eruptions.

Another consequence of lower sulfate levels involves methane. Today, most organic matter decays via oxidation, either directly by oxygen or, once all free oxygen is used up, by a number of other electron acceptors in order of their energy yield, namely nitrate, manganese (IV), iron (III), and, finally, sulfate reduction. This leads to a series of blurry, microbially mediated reaction rings from aerobic to progressively more reduced microenvironments in marine sediment. Because of the high levels of sulfate in the oceans, almost all organic decay today results from a combination of aerobic respiration and microbial sulfate reduction. However, when oceans turn anoxic and sulfate levels are low, different pathways come to the fore, most notably involving methanogenic bacteria, which today are restricted to non-marine environments or the deepest sedimentary layers only. Rob Newton of Leeds University has hypothesized that increased methane flux during times of low sulfate would further exacerbate anoxia by raising the oxygen demand at the seafloor, methane being readily oxidized to carbon dioxide, and driving a positive feedback toward runaway anoxia.

All these events line up in cascading sequence, the key to which is the global sulfur cycle. Beginning with massive gypsum deposition from a high-sulfate ocean, oxygen deficit and anoxia follow, leading to more efficient phosphorus recycling, eutrophication, euxinia, and further expansion of seafloor anoxia. Eventually, following volcanic eruptions, which both warm the planet and accelerate eutrophication via chemical weathering, sulfate levels can become so low as to allow runaway greenhouse conditions and even mass extinctions. Although global climate is largely dependent on greenhouse gas concentrations, there are thought to be many feedbacks that act to mute perturbations to the global carbon cycle. One of the most sensitive of these involves the carbonate system, because the oceans are almost always close to saturation with respect to calcium carbonate minerals, which then precipitate or dissolve depending on the nature of any perturbation. This is not the case for sulfur minerals. Because gypsum is undersaturated in normal seawater, and pyrite formation is dependent on organic productivity as well as sulfate, seawater sulfate concentrations could potentially change within a relatively short interval by as much as two orders of magnitude. Two examples of times when we know that gypsum weathering events were followed by giant deposition events, the Cenozoic Era and the Ediacaran-Cambrian transition, serve to illustrate how such potential elasticity within the sulfur cycle can express itself differently under different earth system conditions.

The Cenozoic Era was a time when physical erosion rates were high due to the collision of India with the rest of Asia, driving the Himalayan mountain range ever skyward. The collision began around fifty million years ago during the Eocene Epoch, but the later Miocene Epoch, which saw giant evaporite deposition, also witnessed major uplift of the Himalayas, as well as during the collisions between Europe and Africa that helped to close the Tethyan seaway. Rates of both pyrite and gypsum weathering would have been high, helping to buffer against cooling forces. The world seems to have been relatively resilient against small perturbations to climate caused by evaporite weathering and deposition, at least compared with times of relatively low erosion rates, such as when supercontinents like Pangaea were reaching their twilight years. Crucially, the highly oxygenated oceans of the Cenozoic Era resisted the spread of euxinia and any deoxygenating effects of oxidative weathering and sulfate deposition. As a result, oxygen and sulfate levels changed only modestly during the Cenozoic Era. Contrast this situation with the Ediacaran-Cambrian transition.

The Ediacaran-Cambrian transition was superficially very similar, as massive mountain building featured heavily during both times, flooding the oceans with sulfate. However, the effects were entirely different in a far less oxygenated world. First, surplus sulfate dramatically changed the redox balance in the anoxic oceans of the Ediacaran Period, causing oxygenation via pyrite burial at the same time as global warming via the sulfur cycle (DOC oxidation) beyond anything that would be achieved during later times via the calcium cycle. High sulfate levels then led to massive gypsum deposition events that dominated after 550 million years ago. High erosion rates consumed oxygen through oxidative weathering, while either euxinia and pyrite burial gave it back (and more) or sulfate deposition removed it, with the result that the spread of anoxia was only a short step away from oxygenation. Global climate, oxygen, and sulfate levels during the Ediacaran-Cambrian transition oscillated elastically through high amplitude cycles mirrored in biological radiations and crises.

From these two analogous tectonic scenarios, separated in time, it is evident that the response of the biosphere to perturbation depends strongly on the state of the earth system at the time, and not just on the nature of the perturbation. Indeed, the same perturbation can even exact opposite responses. For example, in Ediacaran times warming went hand-in-hand with oxygenation, but the opposite was more often the case during later times. If we look back to an even earlier time, we might be able to detect still other responses. I suspect that the Great Oxidation Event (GOE), quite possibly the most significant environmental event in all of earth history, might have been one such moment. Where previously there had been only the merest whiffs of oxygen, in local oases or within microbial mats, now and for all time, oxygen levels would remain consistently above a minimum threshold, sufficient to oxidize the terrestrial weathering environment. A large global project funded by the Alfred P. Sloan Foundation, the Deep Carbon Observatory, has estimated that the GOE produced, amazingly, more than half of the mineral types on earth today. This is because new oxidation states were suddenly made possible for numerous elements. As sulfur was one of those "polyvalent" elements, the sulfur cycle also changed irreversibly, allowing it for the first time to take up the role of System Earth's disrupter-in-chief.[12]

Although no giant evaporite deposits are forming today, calcium sulfate is continually being removed in a different form wherever hot seawater circulates through cracks in the seafloor. This process takes place beneath the volcanic chains, called mid-ocean ridges, where new ocean crust is always forming. We know this because anhydrite, a dehydrated form of calcium sulfate, is a common mineral around vents where pressurized seawater has reached boiling point at temperatures in excess of 150 degrees Celsius. The precipitation of anhydrite clogs up cracks within months to years, drastically reducing the permeability of ocean crust. Because sulfate is almost three times more concentrated in seawater than calcium in today's oceans, one might expect excess sulfate to remain in hydrothermal fluids after anhydrite precipitation. However, calcium is leached from basalts during alteration of the ocean crust in exchange for magnesium, which means that all of the surplus sulfate ends

up being removed. You might think that I should have mentioned all this earlier when I described the effect of sulfate deposition on the global sulfur cycle. However, on long time scales
hydrothermal anhydrite precipitation is usually neglected because most of the anhydrite will eventually dissolve, returning
sulfate back to seawater, once the new ocean crust has cooled
down after moving far enough away from the mid-ocean ridge.
But there must have been a moment in time after the GOE
when this process began. When was that, and would the initi-
ation of such a transient sulfate sink have had a significant effect on the earth's oxygen budget? To answer those questions,
we need to take a slight diversion to before the GOE, when there
was essentially no free oxygen in the surface environment.

Geologists divide earth history into four eons. We have
focused primarily on the latter two, the Proterozoic and Phanerozoic Eons. The earliest, Hadean Eon has left no geological
record and was a time of volcanism and meteorite bombard-
ment on earth. However, the ensuing Archean Eon, which lasted
between about 4.0 and 2.5 billion years ago, seems to have been
the time when the unique features of our home planet first
came into being. Life kicked off during the Archean, for example, and by the time it ended, cyanobacteria were producing
oxygen. This was also the time when the earth's crust evolved
by repeated partial melting into a less dense surface silicate
layer with less magnesium but more potassium. By the end of
the Archean Eon, the earth's low-density continental crust could
simply float on top of the mantle like ice on water. This allowed
denser ocean crust to descend beneath continental crust, leading to the onset of steep subduction, supercontinent cycles,
and plate tectonics as we know them today. The most obvious
effect of buoyancy was to allow continents to emerge above the
waves, turning what had been an almost entirely watery world
into something much more like our present blue, green, and

brown planet. The emergence of rocky continents transformed biogeochemical cycles for all time.[13]

The most obvious change was to weathering. Before the emergence of continents, most chemical weathering would have taken place on or beneath the seafloor. The Archean-Proterozoic boundary is situated toward the end of some of the largest crustal building events in earth history, which led eventually to the formation of the megacontinent Kenora (or Superia), followed by a relative tectonic lull after about 2.3 billion years ago. The huge mountain chains that arose during those events changed the focus of chemical weathering from the marine to the terrestrial realm, where acid rain could now etch away at the freshly exposed silicate rocks, releasing calcium, carbonate, and nutrient phosphorus. The end of the Archean Eon, the Neoarchean Era, witnessed vast carbonate rock platforms, extending from coastal areas to banded iron formations on the deeper basin slopes. The switch to terrestrial weathering allowed oxygen to rise because the silicate weathering carbon sink was now directly linked with the release of nutrients that could drive organic production, carbon burial, and oxygenation. However, the resulting GOE also introduced a new kind of weathering, oxidative weathering, boosted by sulfuric acid from oxidized pyrite, which led to the first surge of riverine sulfate into the ocean.

Reading between the lines here, the Archean-Proterozoic boundary represents a major switch in the long-term carbon cycle. The carbon isotope mass balance tells us that throughout earth history approximately 80% of all carbon flowing through the earth system has been removed as carbonate, with the other 20% removed as organic matter. Before the emergence of continents, most of that carbonate would have precipitated beneath the seafloor in the ocean crust during hydrothermal alteration. In the absence of much silicate, limestone, or oxidative weath

ering, the riverine carbonate flux would have been considerably smaller. After the widespread emergence of continents, the
carbon flux and with it the oxidizing capacity of organic burial
must have become much greater. At the same time, the proportion that silicate weathering made of the total carbon sink
would have become considerably smaller. The ocean crust,
however, would have remained a major carbon sink for as
long as seawater sulfate levels in the deep ocean were too low
for calcium to react with sulfate to form anhydrite. The onset of
hydrothermal anhydrite precipitation was originally believed
to have occurred quite late in earth history because even after
rivers became rich in sulfate, following the GOE, the deep marine environment remained a stubbornly oxygen- and sulfate-
free zone, at least until the Ediacaran Period, while at productive margins euxinic conditions favored the removal of sulfate
as pyrite. All those preconceptions changed, however, when an
international group of scientists drilled the Archean-Proterozoic boundary near Lake Onega, which lies to the northeast of
St. Petersburg in Russia. Amazingly, they found that to reach
the boundary they needed to drill through hundreds of meters
of anhydrite and so discovered the oldest giant evaporite deposit in the world.[14]

What the FAR-DEEP project stumbled across was truly
extraordinary. In some cores, they found more than 500-meter-
thick layers of gypsum, together with halite and carbonate, in
precisely the same mineralization sequence as during the Messinian Salinity Crisis more than two billion years later. By their
calculations, the sulfate concentrations in seawater back then
were at least one-third of modern levels, which is even higher
than has been estimated for global seawater when the southern
Atlantic Ocean opened during the early Cretaceous. The findings supported an earlier, puzzling report of anhydrite in rocks

of similar age from South Africa. Intriguingly, both deposits were found associated with carbonate rocks with extremely high carbon isotope values, indeed some of the highest ever recorded. This isotope event has been named the Lomagundi-Jatuli Event, or LJE, after the regions where it was first discovered in Zimbabwe and Finland, respectively. Clearly, our preconceptions of the early Proterozoic ocean were entirely wrong, and sulfate concentrations in the ocean must have fluctuated tremendously ever since the GOE, with potentially enormous implications for the earth's oxygen budget.[15]

Because sulfate flux into the ocean prior to the GOE would have been very low indeed, most of the ocean's sulfate following the GOE needed to come from the weathering of pyrite, which is a major oxygen sink. For this reason, the oxygen needed for the pyrite weathering and the GOE can only have come from organic carbon burial. If such high sulfate levels did indeed induce the precipitation of anhydrite during the LJE, then any oxygen consumed during pyrite weathering could not return to the surface environment until the anhydrite redissolved millions of years later, once it had reached far enough away from the hot, ocean-spreading ridges. As oceans grew wider, the anhydrite oxygen sink would have grown as well, continuously leaking oxidizing capacity away from the surface environment. Because low sulfate levels returned throughout later earth history, the reintroduction of this transient oxygen sink likely occurred many times over. Massive evaporite deposition during the LJE exacerbated the oxygen leak by depositing sulfate that would also not reemerge via weathering for a geological age. This scenario is the precise opposite of the Ediacaran-aged Shuram anomaly, as is its carbon isotope expression. Instead of sulfate weathering coupled with sulfide deposition, the LJE was associated with sulfide weathering coupled with sulfate deposition. As with the negative Shuram anomaly, the positive LJE anomaly could only be sustained for as long as the imbalance could be sustained.

The LJE is known to have lasted for more than a hundred million years, nestled between about 2.31 and 2.06 billion years ago. It directly followed the GOE and spanned a tectonically uneventful interval with little evidence for the orogeny or magmatism that marked the end of Kenora's tenure as a megacontinent. Just as with Pangaea during the Permian Period, chemical weathering would have been limited by the low rates of denudation on an old, supersized craton. Low weathering rates limit phosphate availability from both continental weathering and ocean recycling, suppressing biological productivity. Such nutrient-starved, or "oligotrophic," conditions hinder bacterial sulfate reduction and euxinia, allowing sulfate levels to rise, eventually fueling gypsum deposition. Counterintuitively, the long tenures of Kenora, Rodinia, and Pangaea all coincide with very high carbon isotope values in carbonates and therefore high proportional amounts of organic carbon burial. However, this does not imply high organic productivity due to the lack of nutrients but is instead related to a combination of low weathering rates and the continuous leak of oxygen from sulfate deposition in the Rhyacian, Tonian, and Permian Periods, respectively. We have seen that several such gypsum deposition events, such as before the Permian-Triassic boundary mass extinction, terminated in eutrophication and euxinia, and the LJE was no exception. Around 2.06 billion years ago, the world's largest layered igneous body, the Bushveld complex, intruded parts of South Africa, just as the Franklin Igneous Province and Siberian Traps would mark the respective breakups of Rodinia and Pangaea more than a billion years later. The orogenies that followed the LJE gave their name to the next geological period, the Orosirian, which lasted between 2.05 and 1.8 billion years ago.[16]

Orogeny brings oxyanions into the ocean, sulfate chief among them. One of the weirdest examples of this increased flux of oxidizing capacity was discovered in Gabon in 1972, at the height of the Cold War. In that year French scientists discovered that rock samples from a uranium mine in Gabon were depleted in the radioactive isotope ^{235}U, which is the same isotope that needs to be concentrated in nuclear reactors. An investigation was launched immediately to ascertain whether uranium had been diverted illegally for the manufacture of a nuclear bomb. It was quickly discovered that the nuclear fission reactions had occurred, not recently in a reactor, but two billion years earlier. Such natural nuclear reactions cannot occur today because most radioactive uranium-235 has now decayed to lead. However, during earlier times, there was a short window of opportunity when oxyanions like sulfate and uranate ions were high in the surface environment, following the GOE. Importantly, the ratio of ^{238}U to ^{235}U was also lower than today, because the radioactive half-life of ^{238}U is much longer than that of ^{235}U, making a potentially volatile mixture. As euxinia became more prevalent at the close of the Rhyacian Period, marked by the end of the LJE anomaly, uranium could be scavenged from seawater in anoxic environments, essentially making a slow-ticking nuclear bomb. If nothing else, the natural fission reactors of Oklo in Gabon are a genuine "smoking gun" for the very first Permian-Triassic boundary-type series of events.[17]

Due to increased sulfate and nutrient flux into a deoxygenated ocean, widespread euxinia and mineral sulfide deposition reigned supreme after the LJE, just as during the much later Shuram anomaly. Some pyrite concretions in the Francevillian strata of Gabon are so large and strangely shaped that they

have even been thought—rather implausibly in my opinion—
to mold the soft parts of large and motile aerobic organisms.
They would be the oldest such fossils in the world if confirmed,
but for the time being the oldest known large organic-walled
fossils, spiral bands of *Grypania,* are from rocks of similar age
in North America. This sequence of events was repeated at least
once more when the supercontinent Nuna finally amalgamated
around 1.6 billion years ago. At this time, gypsum deposits
preceded large organic-walled fronds in North China that are
also associated with a negative carbon isotope anomaly. It may
be that many such anomalies remain to be found, but already
we know of carbon oxidation events from about 2.0, 1.6, 0.95,
and 0.80 billion years ago, as well as numerous similar pulses
of ocean oxygenation both before and after Cryogenian ice
ages and throughout the Ediacaran-Cambrian transition. What
this tells us is that the "boring billion" of carbon isotope stability hides a multitude of perturbations, but essentially confirms that the marine organic carbon reservoir largely held firm
throughout more than 1,600 million years of earth history,
until it all started to go awry just before the onset of Snowball
Earth. Due to the repeated exhaustion of this powerful climate
capacitor, the final 200 million years of its existence were far
from boring.[18]

Ever since the first emergence of continents, the onset of
modern tectonics, and the arrival of an oxygenated atmosphere,
supercontinent cycles have held sway over oxygen budgets in
our surface environment with predictable consequences for the
biosphere. Opportunistic radiations and extinctions of aerobic
organisms have taken place not only since the Cambrian explosion but throughout Proterozoic earth history as well. Similar tectonic and environmental events have also led to rather
different consequences, as the oxidizing capacity of the surface
environment has increased over time and ecosystems have also

evolved first multicellularity and then complexity. Among these events, the fate of calcium sulfate, whether it ends up as gypsum or pyrite, turns out to be the wild card, throwing the system out of kilter one way or the other. I don't think we can speak any longer of earth history as a series of step-like changes. Ever since oxygen first rose in the atmosphere, the biosphere has been at the mercy of tremendous tectonic forces that store oxygen in a gypsum battery only to deliver it wholesale to an unsuspecting world millions of years later, causing extremes in both oxygen and climate.

Snowball Earth and the Permian-Triassic mass extinction, first identified by John Phillips, are now joined by the extraordinary events of the Rhyacian-Orosirian boundary, all similar in origin but each with its own set of earth system boundary conditions and very different consequences for the evolving biosphere. Earth history lays bare the interplay between time's arrow and time's cycle, but how inevitable is the evolutionary trajectory that results? Can we perceive any guiding hand here, steering conditions toward the Goldilocks environments conducive to life? Using the metaphor of Stephen Jay Gould, if we replayed the tape of life, would we end up in the same place as today, with a fully oxygenated surface environment and intelligent life such as ourselves? In a recent study, Toby Tyrrell, a chemical oceanographer based at the University of Southampton, put 100,000 randomly generated planets through their paces, with the result that only 8.7% remained habitable in at least 1% of the trials. Does this tiny success rate mean that we earthlings are simply fortunate to have evolved on a rare planet, blessed with more than its fair share of luck? A question worthy, perhaps, of a heated discussion in London's Freemasons' Tavern or while strolling along Darwin's thinking path. In the final chapter, I would like to moot some tentative answers to this question.[19]

12

Time's Arrow

Glaciation provided a conducive environment for animals to evolve. Volcanism boosted the earth's greenhouse blanket, melting the snowball. Ensuing weathering fertilized the oceans with nutrients and oxygen that weaponized an arms race: the Cambrian explosion. Extinctions and radiations created bottlenecks and opportunities for all of our ancestors through to the present. If life's progress was contingent on such past events, does that mean that the emergence of complex life was vanishingly improbable both here on earth and on similar planets elsewhere? Gaia theory predicts that once life evolves, it hones the environment to its advantage, but without external forces, evolution would have no direction. Although animal evolution depended on chance innovations on the road to complexity, notably the genetic machinery required for oxygenic photosynthesis and the implausible symbioses necessary for eukaryotic life, extrinsic forcing mechanisms like the sun and plate tectonics predetermined life's march toward larger and more energetic forms. Ice may have helped to navigate us here, but in the driver's seat was primordial Fire.

I t is the summer of 2019, and James Lovelock is celebrating his one hundredth birthday at the University of Exeter with a symposium and a suitably enormous cake. There is music, poetry, meditation, and reminiscence, as well as science at this most eclectic of conferences. In a typically no-nonsense interview with Tim Lenton, probably the best-known advocate of Lovelock's ideas among today's crop of earth system scientists, "Jim," as we all seem to call him, regales us with tales from a lifetime of insatiable curiosity and invention, much of which preceded his most famous contribution to science: Gaia theory. Despite substantive critique, Gaia has stood the test of time. As metaphor, inspiration, and hypothesis, it still excites debate, fully half a century after Jim began formulating the hypothesis with the biologist Lynn Margulis. Part of the reason for its popularity is that Gaia theory proposes a partial solution to the eternal quandary of how we came to be here. For those, like me, who are fascinated by our origins, Gaia is a constant source of inspiration.[1]

While launching myself into my studies in Zurich, I decided that I needed to write to James Lovelock. At the time he lived as an independent scientist at his "experimental station" not far from the English town of Plymouth, where I grew up. His generous reply took my naive inquiry seriously and introduced me to a fascinating new idea. Jim had recently heard that Don Anderson, a renowned geophysicist from the California Institute of Technology and author of the book *New Theory of the Earth*, was pondering whether without life the earth's liquid oceans would have evaporated away long ago due to our warming sun. This is because life reduces the greenhouse effect by removing carbon dioxide from the atmosphere. Organisms do this first by squirreling away carbon dioxide in the form of fossil carbon and second by catalyzing chemical weathering, which supplies oceans with the nutrients and ions that drive both organic and carbonate carbon burial. Geologists have long argued that without liquid water to lubricate subduction, plate tectonics would quickly cease, and so, Don mused, even Hutton's rock cycle might owe its existence to life on earth.[2]

Science moves forward by the acquisition of new data to put such crazy notions to the test. We know more today than we ever did before, indebted to the curiosity of those who preceded us. Indeed, the story I have told thus far is based on both wild ideas and new data that simply did not exist a mere generation ago. Although the evolution of human knowledge is undoubtedly contingent on chance genius and lucky discoveries, its overall trajectory is far from random, despite occasional false leads and setbacks. Knowledge has a direction. It is not for nothing that we use the term scientific *progress*. Indeed, we take continual improvement so much for granted that advertisers easily convince us that any new technology will be better than what came before. Perhaps this is why Ying and I keep buying bagless robot vacuum cleaners that merely push balls of dog hair around the house rather than suck them up.

Biological evolution shows the same dependence on chance and contingency, but, despite suffering such apparent setbacks as Snowball Earth and mass extinctions, it has, like knowledge, continued stubbornly in one direction toward greater complexity. How is it that life has shown such mindless "progress"? Is it due to a succession of ludicrously implausible coincidences, helping us to win the universe's biggest lottery again and again? Or was the evolution of complex, energetic animals a predictable outcome on a living planet? In other words, would it happen again, given the right starting point? And if so, can we attribute each gentle nudge toward modernity to the largesse of Mother Gaia or to the earth's favorable size and distance from the sun? As Goldilocks might say, our home planet might simply have been not too hot, not too cold, but just right.

Gaia theory proposes that life, once initiated, helped to create the conditions conducive to its own success, resembling how algae and fungi combine in the form of a lichen, cooperating in symbiosis to their mutual advantage. Lynn Margulis, best known for championing the symbiotic theory of eukaryote evolution, was unsurprisingly the co-founder of the Gaia hypothesis. Although early versions were widely ridiculed at the time, her apparently unhinged idea that eukaryotes evolved when certain types of bacteria began cooperating inside other microbes, later discovered to be archaea, is now the conventional view. Those first parasitic bacteria lost most of their DNA over time, eventually becoming the mitochondria that power our bodies, entirely dependent on their host and trapped inside its cell walls. Scientists are now equally certain that chloroplasts in plants and algae began similarly, as symbiotic cyanobacteria, just as Margulis predicted. Symbioses created some of the most important biological innovations needed for complexity, but there is some fogginess over how there could be a symbiosis between life and its inanimate environment. After all, life undoubtedly needs the cradling comfort of its planetary environment, just as the bacterial symbionts that became mitochondria needed their cellular host. However, the reverse cannot be said of the planet, which hardly needs life to survive. Still, if connections can be found between life and its environment, in both directions, then this lends credence to the idea of Gaia as a self-regulating, or homeostatic, entity. Symbionts originate when one organism's garbage is another's food. Such cozy or, depending on your viewpoint, sinister interdependence is the stuff of life.[3]

One of the first criticisms of the Gaia hypothesis was that environmental regulation would not be an emergent property of life because organisms are by nature competitive, mindlessly draining resources with abandon. Lovelock and his erstwhile

doctoral student Andy Watson, now himself a renowned fellow of the Royal Society, invented the computer model Daisy-
world to refute the critique. Their virtual experiment, in which
competition between pale and dark-pigmented daisies regulates planetary temperature, led to a reformulation of Gaia
wherein life equates to negative feedbacks. Watson's doctoral
study of the role that fire plays in regulating oxygen levels is
just such a negative feedback. Although wildfires began with
the first forests of the Devonian Period, analogous oxidation
reactions have undoubtedly helped to regulate oxygen levels
ever since the Great Oxidation Event (GOE). However, life has
persisted through all manner of extremes at the earth's surface,
and still exists under extraordinarily varied conditions today.
Clearly, no oxygen level can be ideal for all life. Moreover, the
amount of free oxygen in our atmosphere and oceans does not
relate directly to organic production but instead to the extent
to which oxygen sinks counteract the oxygen released when
dead organic matter escapes those sinks. As a consequence, the
idea that life "governs" its redox environment is not entirely
convincing; perhaps feedbacks are simply part and parcel of all
complex systems like the earth.[4]

Lovelock recalled during his centenary celebration that it
was his work for NASA that started him thinking about Gaia.
During the search for life on Mars, he noted that the atmosphere
of Mars made sense in terms of abiotic chemistry, unlike the
earth's. In other words, on Mars there had been sufficient time
for dynamic equilibrium to be established by, for example, the
transfer of electrons between rocks and the environment to reach
an energetically stagnant, more stable state. Chemical equilibrium, he suggested, was a fundamental distinguishing feature of
a dead versus a living planet. A chemist's term for the transfer
of electrons is oxidation (and its counterpart reduction), with
one of the most energy-releasing oxidation reactions being the

direct addition of oxygen to any "more reduced" substance willing to loosen its grip on an electron or two. The evolution of a new type of cyanobacterium that favored oxygen as the terminal electron acceptor thus created a highly reactive chemical species out of equilibrium with its ambient environment. Free oxygen not only fuels highly energetic metabolisms, such as our own respiration, but enables a cascading series of microbial ecosystems that exploit the chemical potential between reduced organic matter and its more oxidized environment. Once the oceans became more fully oxygenated during the Ediacaran Period, the surface area separating contrasting energy regimes, and with it the earth's biomass, expanded enormously not only in sediments and the water column, but even inside the bodies of newly evolved animals, which had evolved into fabulously fractal forms that maximized contact between organic membranes and ambient oxygen.

We animals are heterotrophs, which means that we get our energy by oxidizing organic compounds rather than from more direct sources like the sun. However, the reason that we are able to do this is because reduction-oxidation, or redox reactions, occur spontaneously in our highly oxidizing environment. Like all organisms, we get our energy by exploiting an emergent energy gradient, caused ultimately by the separation of forward and backward reactions. To see why this is the case, let's follow how organic matter is broken down. Imagine the fate of an organic parcel, be it a dead organism or some fresh excrement, descending to the seafloor, or some food making its way through our gut. Once there, if oxygen is available, aerobic respiration is easily the quickest decomposition process because it releases the most energy. Benthic animals and their churning up of the sediment catalyze this process. However, even in their absence, the urge for reduced organic carbon to decay to carbon dioxide in a more oxidized environment would

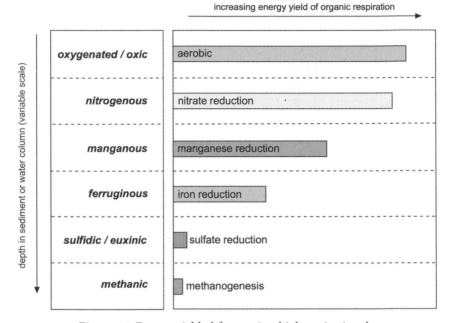

Figure 26. Energy yielded from microbial respiration, by
different types of organic decomposition (respiration) reactions.
If environments are sufficiently oxidized, then vastly more energy
can be released during organic decay, driving a chain of microbially
catalyzed reactions as oxidants (electron acceptors) become
consumed. The dominant metabolism of microbial communities is
dictated by energy yield, leading to chemical and microbial tiering
in the natural environment.

guarantee the same result, albeit a little more slowly. Once all
oxygen has been consumed, and the environment turns anoxic, other microbial metabolisms compete over the remaining scraps, in an order determined by how much energy would
be released. First, all available nitrate is consumed, followed by
iron (III) and other transition metals, then sulfate, which gets
reduced to sulfide, and finally barely viable methanogenesis
and fermentation (Figure 26).

In a refutation of Gaia, Peter Ward has invented her nemesis, Medea, a goddess, whose power is as destabilizing as Gaia's is harmonious. Ward is a prolific paleobiologist and science writer who has enjoyed a long association with the University of Washington in his birthplace, Seattle. According to Ward, Medea takes the upper hand when conditions become more anoxic, and therefore less favorable to animal life. The toxic byproducts of anaerobic microbes, such as hydrogen sulfide, make life increasingly unbearable for animals, leading to destructive positive feedbacks as the microbes try to seize back control of the planet. As we saw already, however, bacterial sulfate reduction was key to Ediacaran oxygenation, and needs a continual source of sulfate, which means Medea may be little more than a dramatic metaphor. But is Gaia any different?

In his book *Almost Like a Whale,* Steve Jones describes evolution in fabulously pithy terms. "Common sense," he writes, "tells us that life—like the Sun—revolves around ourselves. The idea has but one fault. It is wrong." Gaia theory may suffer from a similar problem, in that it would be impossible for us to live on a planet that did not support our existence as well as the existence of all of our ancestors. Indeed, for you and me to be here on this planet, there must have been an unbroken line from each one of us back to some mutual ancestor and then just one direct line to some pond scum four billion years ago. In other words, Gaia theory could be a version of the logically flawed but highly attractive "anthropic principle." Instead of giving life the credit for conditions that help maintain and perpetuate its own existence, we may simply be lucky enough to live on a special planet, on which conditions never proved lethal to *all* life and have become increasingly conducive to *human* life. As we are among the latest in a long series of organisms to evolve, it is unsurprising that our highly energetic metabolisms are best suited to today's oxygenated world, in which the energy yield of aerobic respiration has reached its all-time peak. By contrast, anaerobic bacteria have long since been marginalized to hidden populations in swamps and sediments, inside our bodies, or even inside our cells just to survive. However, that does not mean that anaerobes are vulnerable, as they thrive on our excrement down the chain and are probably doing as well as ever, if not better. Although life covers a range of different organisms with a range of different requirements, does the tale in this book tell us anything about the potentially stabilizing effects of the living world? Or does life always exploit opportunities conjured up by events entirely beyond its control?[5]

Critics of Gaia point out that destabilizing positive feedbacks exist alongside negative feedbacks, and we may just be lucky to live on a planet that is currently dominated by the ship-steadying negative variety. Toby Tyrrell and David Waltham, leading opponents of Gaia theory, have both written thoughtful books that are critical of our tendency to overplay the importance of negative feedbacks, such as the silicate weathering thermostat, pointing to how events like mass extinctions demonstrate that our biosphere can shift from one "normal" to another with catastrophic consequences for some forms of life on earth. The Medea hypothesis suggests that such events are attempts by microbes to take over the world, and it is of course true that life itself, as opposed to merely animal life, did not even come close to extinction during so-called biotic crises. However, if life is defined as all life, including microbes, then Gaia becomes a very slippery fish, reducing almost to the point at which we began this chapter. In other words, all that remains of the hypothesis is whether the cycling of liquid water and essential nutrients, which are the most fundamental requirements of *all* life, is actively perpetuated by *any* life.[6]

Snowball Earth is perhaps the best example of a fundamental earth system shift to a new normal state, as it lasted for so incredibly long. Although extreme by current standards, the Cryogenian glaciations were essential to the further evolution of animals, while the true biological catastrophe likely occurred afterward when life forms that had adapted to permanent sub-zero temperatures were suddenly subject to the opposite end of the climate spectrum. Clearly, life survived the Cryogenian deep freeze, but we can hardly describe the planet at that time as providing ideal conditions for life, as energetic metabolisms would have slowed down to barely ticking over. Moreover, the way out from global glaciation required volcanism, which lacks any obvious biological control, unless again you take the extreme view that tectonics can only be sustained on a living planet. Life has survived a huge range of temperatures, salinity, aridity, and atmospheric composition without once becoming extinct or even reaching a major dead end on the path toward more energetic metabolisms and increasing complexity in both form and function. All animal body plans, or phyla, that evolved during the Cambrian radiations are still with us today, while even our own phylum, the vertebrates, started out during those heady times of wild environmental fluctuations. The concept of climate regulation by life begins to fall apart when we view the way life tends to adapt to extreme events rather than shape them. This is not to say that life played no role in earth system changes, because it clearly does affect its environment, but that life cannot be said to have created the climatic conditions conducive to its own success. So much for climate, but does life regulate oxygen?

The Great Oxidation Event changed our planet beyond recognition. Without cyanobacteria producing oxygen, the GOE would surely never have happened. This much is a given, although without the recycling of rocks and nutrients by tectonics, anything remotely resembling the GOE is hard to envisage.

Life, according to the biochemist Albert Szent-Györgyi, is nothing but an electron looking for a place to rest. In the case of photosynthesis, the electron is stripped away from its donor by the energy of the sun and given to sulfur, iron, or in the case of oxygenic photosynthesis, oxygen in water. Photosynthesis breaks apart the water molecule, releasing oxygen that can only remain free as long as organic matter stays buried in rocks. Therefore, although organisms are responsible for oxygen production, environmental oxygenation needs the rock cycle, which represents the ultimate way in which life-giving energy gradients are produced by the separation of forward and backward reactions. Moreover, there are plenty of natural ways to sap free oxygen from the atmosphere, so it seems unlikely that the GOE could have occurred until the source of oxygen (organic burial) was able to overcome its sinks (volcanism, weathering). As we have seen, organic production requires nutrients that derive from chemical weathering of the newly buoyant continents toward the end of the Archean Eon, which means that the permanent oxygenation of the earth's surface environment, like the governance of climate and salinity, ultimately had a tectonic origin. It seems yet again that the only refuge for a deterministic Gaia would be if tectonics itself were a result of life.[7]

As mentioned in an earlier section of the book, the aerobic respiration of animals and their bioturbating actions actually reduced the amount of free oxygen in the environment, which could help to explain why it took a further 200 million years after the Cambrian explosion for atmospheric oxygen to recover even to Ediacaran levels. On the other hand, we have also seen how animals package organic matter in various ways and that extra organic ballast would have moved oxygen sinks to the seafloor. In this way, animals act as the oceans' vacuum cleaners. But ventilating the marine environment is merely redistribution and does not explain the inexorable oxidation of the earth's surface environment. On a longer time frame, tectonic redistribution of gypsum salt explains the timing of oxygenation or deoxygenation events, but there must also be something else at play here, and once again we need to turn to tectonics, with a little help from the botanical world. There are good reasons why our planet has more free oxygen now than at any time before in earth history.[8]

Because the source of all oxygen is photosynthesis coupled with sedimentary burial, the amount of oxygen that can be released is determined by carbon dioxide, which derives ultimately from the deep interior of the planet. Although most carbon dioxide is simply recycled by tectonics, some has rarely, if ever, seen the light of day before, having traveled from the innermost reaches of the mantle, delivering fresh oxidizing capacity to the earth's surface. Multiply that primordial outgassing by the age of the earth and you get some idea of why organic production today can release more oxygen to the surface environment than ever before. One important point that is easy to overlook is that primordial carbon dioxide does not effuse from volcanic arcs above subduction zones, but instead at ocean islands like Hawai'i and the Seychelles that overlie hot spots near the core-mantle boundary. Such deep-seated magmatism was presumably more common in the early days when life first evolved. Nevertheless, the key to retaining that oxidizing capacity lies in the rock cycle, which, although very long, is still much shorter than mantle mixing times. Courtesy of tectonics, increasing amounts of carbon have cycled through the surface environment, while, conversely, atmospheric carbon dioxide levels have been decreasing. Key to this counterintuitive decrease has been chemical weathering, which has increased in tandem with the sun's gradual warming and with the help of evolving soil biota. It is tectonics that provides the material to be weathered but also the nutrients needed to sustain carbon burial. Over time, tectonic processes have also helped to change carbonate into carbon dioxide through metamorphic decarbonation in subduction zones, while increased continental freeboard has helped shift weathering from the seafloor to the continents. All of these changes have helped drive the recycling of ever more carbon dioxide through the earth's surface environment.[9]

All roads seem to lead to tectonics. Many geologists have concluded that plate tectonics requires the lubricating effect of liquid water. Thus, we return full circle to the question of whether life was ultimately responsible for the onset and perpetuation of plate tectonics, and therefore also of life (and Gaia). It has been estimated that without life's influence, carbon dioxide levels would be ten to one hundred times higher than they are today, risking partial evaporation of the oceans at greater than 70 degrees Celsius and turning the biosphere into a pressure cooker. This is to say nothing of the potential effect that life had in reducing atmospheric pressure through nitrogen fixation and subsequent storage of nitrogen in the earth's mantle, or in changing albedo and cloud cover through evapotranspiration. However, most geologists consider the onset of tectonics to have been a consequence of the earth's cooling interior. Once a mass of molten silicate rock, our planet has been slowly freezing inward. The outer core is still molten from when a rogue planet smashed into the proto-earth, forming the moon. As the convecting outer shells of our planet cooled, thermal gradients at its outer edge shallowed, causing the brittle lithosphere to become denser and thicker, paving the way for deep subduction and modern plate tectonics during the Archean-Proterozoic transition. Tectonics then led to the first high mountain ranges and a modern-looking rock cycle. According to Robert Stern and Taras Gerya, co-conspirators on a new International Lithosphere Program focused on what they call "Biogeodynamics," the cooling of the earth represents another far-from-equilibrium process like the redox gradients that drive organic oxidation. They predict that tectonics is simply an emergent phenomenon to be expected on any cooling silicate planet of the earth's size.[10]

A complementary idea discussed in Nick Lane's book *The Vital Question* describes how the natural heat and energy gradients at the interface between the earth's hot interior and the seafloor provided the perfect environment for the evolution of life itself. In his book, Nick, a biochemist with an astonishingly bright and critical mind that is admirably open to ideas from outside his own field of expertise, addresses the question of why living cells use redox chemistry as their source of energy. He concludes that life evolved on the seafloor due to the energy gradients produced when silicate minerals, once comfortable in the earth's hot, pressurized, and reducing interior, found themselves out of equilibrium in the cooler, less pressurized, and more oxidizing surface environment. Nick makes the point that the reactions needed by those very first cells would have arisen spontaneously due to the natural flow of energy at redox boundaries. As he writes, "These reactions . . . are the only way of dissipating the unstable disequilibrium of reduced, hydrogen-rich alkaline fluids entering an oxidized, acidic, metal-rich ocean." In other words, life began as a "side reaction of a main energy-releasing mechanism": all that was needed were rocks, liquid water, and plenty of carbon dioxide. Once again, it is impossible to consider the organism or its metabolism apart from its permissive environment.[11]

The process by which hydrogen is released from water interacting with rocks is called serpentinization. Serpentine is a soft, soapy, hydrated variety of magnesium silicate, utterly unlike its denser, dehydrated precursor mineral olivine. Serpentinization occurs spontaneously when mantle rocks meet the ocean, releasing heat in the process. As well as being a possible crucible for life, the weakening of rocks due to serpentinization also acts as a vital lubricant for plate tectonics, which therefore ought to be ubiquitous in the universe wherever earth-sized silicate planets occur within the Goldilocks window where water condenses to form liquid oceans. Far from life permitting tectonics, it seems likely that happenstance allowed both life and tectonics to initiate and co-evolve due to the natural energy gradients at the interface between solid rock and liquid water at the bottom of the sea.

Thus began the effective separation, in both a redox and an energy sense, between the earth's surface and interior environments. Some of the hydrogen produced by serpentinization would have been lost to space in those early days, permanently oxidizing it. Not only that, but organic compounds produced by the earliest abiotic and biotic steps to LUCA, life's universal common ancestor, would have fossilized in nascent rock cycles, storing reducing power and cementing the redox gradient between those two worlds. With time, the sun must have taken over from the silicate earth as the energy provider, as we know from biochemistry that photosynthesis derived from respiration, rather than the other way around, leading to intensified cycling and gradients. Since those early days, the co-evolution of life and its environment has been a story of balance between the reduction (by life) and oxidation (by the environment) of carbon, and subsequently of other vital elements like sulfur and iron, at the surface, together with the storage of those same compounds within the rock cycle.

In recent times, it has become increasingly evident that liquid water was present on earth even during the early years of the Hadean Eon, long before life could be sustained, as well as throughout the ensuing Archean Eon. The much smaller amounts of energy supplied by anaerobic metabolisms before the GOE imply a much smaller biomass that seems unlikely to have been able to influence climate, especially as carbon dioxide levels needed to be much higher back then to counteract Carl Sagan's Faint Young Sun. Indeed, modeling studies suggest that abiotic silicate weathering on the Hadean earth kept temperatures at or below zero—when the planet was not being bombarded by asteroids or scorched by volcanic magma, that is. This contradicts earlier assumptions of very hot early surface oceans that have become cooler over time. If liquid water was truly the key to early crustal formation and ultimately to plate tectonics, buoyant supercontinents, uplift, erosion, and the entire rock cycle, as well as to life, then this would be a strong argument in favor of a lucky silicate planet formed within the Goldilocks zone of the solar system. The main prerequisite for the GOE may therefore simply be silicate earth's distance from the sun, which permitted liquid water, with both tectonics and life as entirely predictable emergent properties.[12]

The tectonically triggered GOE heralded a new world in which tectonics, again, could upset the equilibrium via the sulfur cycle. The GOE helped to store oxidizing capacity, originally from photosynthesis, in the form of oxyanions like sulfate that could build up over long intervals of tectonic stasis, during which supercontinents denuded to flat deserts. This happened to the megacontinent Kenora and the supercontinents Nuna, Rodinia, and Pangaea in turn, at about 2.1, 1.6, 0.7, and 0.3 billion years ago, respectively, when high sulfate levels permitted the formation of giant gypsum deposits that later released their oxidizing capacity during times of uplift and erosion. Life was instrumental in those events as continental weathering catalyzed organic production that drove microbial sulfate reduction. And photosynthetic life forms responded to those changes by evolving mechanisms to reduce oxygen toxicity, in the form of heterocysts or carbon concentration mechanisms. Indeed, even the first eukaryotes and algae may have evolved to protect the bacterial way of life in the presence of increasing amounts of reactive free oxygen. Although life was evidently a catalyst of these grand biogeochemical cycles, the overriding triggers of change have been abiotic tectonic events. Life seems to have been riding on the coattails of much larger phenomena.

There is, however, one fundamental way in which life leads to stabilizing negative feedbacks, although the primary driver of this process is the sun. Solar energy provides the means to reduce carbon dioxide to a fuel, thus driving other metabolisms, like ours, that can exploit the energy differential between reduced carbon and the more oxidized environment. As time has progressed, metabolic reactions have become increasingly energetic and widespread, releasing tremendous amounts of stored solar energy. Although life seizes on resources with abandon, the propensity for metabolisms to use both reactant and product of redox reactions leads to inevitable negative feedbacks as exploitation of one leads to impoverishment of the other. Like the wild swings in populations of foxes and rabbits (and grass) on a hypothetical desert island, such games of tug-of-war are the stuff of life. Chemists would recognize this as an example of Le Châtelier's famous principle, whereby products and reactants re-equilibrate following perturbations. Although positive feedbacks destabilize the environment at times, it seems that, short of total extinction, eventual domination by negative feedbacks is an inevitable consequence of life, not in spite of, but because of, perpetual competition for resources. It is perhaps this tussle that has ensured that Ward's microbial Medea has never won her battle to destroy complex life. That this leads to ideal conditions seems unlikely, however, as life forms compete for different resources under different circumstances.

An abiotic planet such as Mars is not devoid of redox extremes at its surface, and is also subject to solar forcing and an internal heat engine, but the existence of life on earth sharpens energy gradients, maximizing their potency, while tectonics continually supplies and stores both fuel and accelerant in abundance. It spins my head to give life credit for both stability and variability through time and space. But if we are to reject life-giving homeostasis as the reason for directionality in evolution, of both life and the planet, what are we left with? Sheer dumb luck?

I began this book, as I also began my geological mapping of Islay in 1990, looking at glacial deposits of the Cryogenian Period. When I began that fieldwork in Scotland as an undergraduate, the idea that our planet had once been entirely covered in ice was still considered fanciful. Now, thirty years later, not only is this theory broadly accepted, but we also know for how enormously long glaciation held our planet in its icy grasp. The Sturtian ice age alone lasted for as long as there have been primates on earth. The fact that both glaciations were preceded by considerable perturbations to the global carbon isotope mass balance confirms that tectonically driven but microbially catalyzed oxidation was a key player in rendering the earth vulnerable to climate catastrophe. It is true that an oxygenating seafloor proved a perfect hatchery for larger, benthic, aerobic organisms, and we see such organisms radiating opportunistically as early as two billion years ago. However, over more than a billion and a half years, these life forms disappeared each time as abruptly as they appeared and could not establish themselves without a permissive environment. Indeed, eukaryotes as a whole seem to have played second fiddle to microbes right up to the Ediacaran-Cambrian transition. The world's oceans stubbornly resisted oxygenation for almost all of earth history, and it was only when tectonic events finally delivered enough stored sulfate to eradicate the DOC capacitor that entire oceans could become pervasively oxygenated. The first two occasions led to global glaciation, while the later Ediacaran and Cambrian episodes were transient peaks in cycles from one extreme to the other. Aerobic organisms, such as animals, seem to have been at the whim of tremendous fluctuations in habitable ecospace, leading to the evolutionary bottlenecks and radiations that are the hallmark of the Cambrian explosion. There was no step change from a primitive to a modern earth system, either climatically, environmentally, or even biologically, but instead a noisy transition through which our direct ancestors' line scraped through unscathed.

Part of the problem in maintaining the oxygenated conditions conducive for complex life is that biogeochemical cycles take with one hand while giving with the other. Just as organic burial produces life-giving oxygen today, it also generates the oxygen sink or fuel of tomorrow. Most of the pyrite buried during the Shuram anomaly has now consumed the oxygen it once released by being weathered, and it is the same for organic matter. Perhaps once, a large proportion of organic carbon was simply recycled as detritus, but in today's more oxygenated environment, almost all fossil organic matter will be oxidized and so represents an enormous oxygen sink. How is it that this zero-sum game has led to any change at all? Undoubtedly, we live during the most oxygenated time in earth history, with record amounts of oxygen in the atmosphere, sulfate in the oceans, and ferric iron in the earth's crust. How did these perfect conditions for our own evolutionary success come about?

The story of the oxidizing earth system is one of an accelerating global carbon cycle with higher outgassing, higher organic production, and higher nutrient availability, but against the backdrop of an ever shrinking surface reservoir of carbon dioxide. High flux and a small reservoir is a recipe for vulnerability to perturbation, yet we haven't seen anything like the extreme changes of the past, which suggests that our biosphere today is fairly resilient against change and that negative feedbacks prevail. One of these feedbacks, the carbonate compensation depth, or CCD, in the ocean, appeared only in the Mesozoic Era, since which time ocean anoxic events have not led to the kind of mass extinctions seen earlier in earth history. The CCD is the depth in the ocean beyond which carbonate no longer appears on the seafloor because the oceans are undersaturated with respect to calcium carbonate minerals due to a combination of low temperatures, high pressures, and fluctuating pH. The CCD is a result of the evolution of pelagic calcifying plankton, and so did not exist in any meaningful sense as a capacitor before such organisms evolved during the Mesozoic Era. It is today one of the most important negative feedbacks counteracting changes to atmospheric carbon dioxide levels on time scales of hundreds to thousands of years due to the length of time it takes to mix the oceans.

We can look at this is in a different way using the carbon mass balance. In Archean times when continents were largely submerged, weathering feedbacks were fairly inefficient, being largely submarine in the case of carbonate burial, while buried organic matter simply did a rinse cycle due to the lack of free oxygen. Today, most of the carbon flux is from weathering, whereby carbonate weathering has little long-term impact on carbon dioxide levels, although metamorphic reactions can extract the carbon dioxide back from buried marble. The net carbon sink today is dominated by a combination of organic burial and carbonate burial coupled with silicate weathering, and both of these processes are tightly linked to chemical weathering. The proportion that organic burial makes up of the net carbon sink, that part of the carbon cycle comprising only outgassed carbon dioxide, has changed from one-fifth to three-fifths. This means that about 20% of outgassed carbon dioxide from volcanism ended up as organic matter when continental weathering was limited in the Archean waterworld. Over time, this portion has risen nearer to 60%, despite the much larger short- and long-term carbon flux.[13]

This huge change is testimony to both the greater efficiency of the organic carbon cycle today, thanks to the continual evolution of carbon concentration mechanisms, and the preferential release of nutrient phosphorus from rocks during weathering with the help of land plants. Carbon concentration mechanisms have evolved over time, on account of rises in oxygen relative to carbon dioxide partial pressures, in photosynthetic organisms from cyanobacteria after the GOE to C4 plants like grasses that radiated during the descent into glaciation from the Oligocene Epoch to the Miocene. Add to this the cycling of carbon through the surface environment, with each carbon atom being metabolized at least a thousand times before burial, in large part due to its reactivity in a more oxidizing environment, and we have the recipe for tremendously responsive feedback. If Gaia has stood the test of time, then today's world, and probably the future biosphere, too, must correspond to peak Gaia, as life is so intimately involved in catalyzing all major biogeochemical cycles today. Gaia may eventually perpetuate habitability by helping to keep the earth cool. Moreover, strengthened stabilizing feedbacks may be an emergent property of life on the silicate earth, even though things did not start out that way.[14]

The story of life's evolution is one of increasingly energetic (or exergonic) reactions due to the inexorable rise in free oxidizing capacity. We can see this in the evolution of biogeochemical cycles, which have progressively developed new oxidative pathways made possible by the energy gains across new redox thresholds in an expanding number of new environments.

Events like Snowball Earth and mass extinctions were merely bumps on the road toward an ever more widely oxygenated world where exergonic metabolisms, such as our own, could thrive, banishing cascades of less energetic, microbial metabolisms beneath the surface. Although tectonic cycles have determined the timing of major events, time's arrow toward our high-energy world has never faltered because two aspects, the strengthening of the sun's rays and the cooling of the planet, have been inexorable forces. To imagine that life keeps the earth habitable seems to be at odds with the notion that the major driving forces are far outside life's control. In other words, once life begins it is very hard to shake off, and so becomes an intimate part of the earth system, but that does not mean that it has ever been in the driver's seat.

Taking a step back from specific geological and environmental events, we can see that energy flow has been the constant driver of change, to which life has continually adapted to exploit the natural energy gradients related to solar flux, mantle heat loss, and the vagaries of tectonic cycling. Metabolic reactions resemble the eddies on the edges of waterfalls. Emergent oases of low entropy, eddies cannot exist without their exergonic parent, the waterfall. The waterfall, on the other hand, is entirely oblivious to the existence of its more exquisitely structured offspring. And so it is with life. Although only 0.3% of solar radiation translates into the chemical energy needed to produce sugars in gross primary production, which then cycles up to a thousand times due to today's highly oxidized environment, huge amounts of energy still get stored over billions of years. The 264 terawatts of energy taken up by life is considerably larger than the earth's internal heat engine. For example, the energy that powers mantle convection and crustal cycling is more than an order of magnitude lower. According to the second law of thermodynamics, all processes have a natural

directionality determined by their energy states. A hot cup of
tea cools down, but does not spontaneously heat up again. Leaves
go brown and disintegrate, but without energy from the sun,
they do not re-form spontaneously from their constituents. For
four and a half billion years, the earth has bathed continuously
in solar energy, which life has stored in organic matter and
pyrite. Volcanic outgassing and an increasingly hot sun have
ensured that ever greater amounts of this energy could be
stored over time, maximizing the available potential energy
when hydrocarbons or other reduced species reemerge into
an increasingly oxidized environment. The key here, as with
the biochemistry of our own cells, is to separate reactants from
products and in so doing promote naturally exergonic reactions. Otherwise, we are left with a dynamically exchanging,
but compositionally stagnant soup of ingredients.[15]

There has been no monotonic increase in oxygen because
supercontinent cycles are continuously tilting the redox balance between reduced and oxidized compounds. When either
has been in ascendance, different metabolisms have taken advantage, sometimes driving positive feedbacks toward a new
normal, but never altering the overall direction of travel. Animals may have arisen from ice and thrived due to the oxygen
from gypsum salt, but the evolution of increasingly energetic
metabolisms and a larger, more varied biosphere was sustained
by the increasing amounts of carbon dioxide released from the
fire of volcanoes and the increasing amounts of oxidizing
power fueled by the fire of the sun. If you can handle just one
more pinch of salt, you might care to ponder with me whether
our presence on this modestly sized silicate planet might just
be a fairly predictable consequence of our Goldilocks position
in the solar system. I marvel over what this might tell us about
the likelihood of complex, energetic, and possibly even intelligent life elsewhere in the universe.

Notes

1
Time Travel

1. The outstretched arm metaphor is from John Mcphee, who described it in his book *Basin and Range* (1981), the first book in a series titled Annals of the Former World, published by Farrar, Straus and Giroux.

2. The time scale shown in Figure 1 is modified from the internationally agreed-upon geological time scale as shown in the book *Geologic Time Scale, 2020*, edited by F. M. Gradstein, J. G. Ogg, M. D. Schmitz, G. M. Ogg (Elsevier Science Limited). This proposed version is published in Shields, G. A., et al., "A template for an improved rock-based subdivision of the pre-Cryogenian timescale," *Journal of the Geological Society* 179 (2021). A template for an improved rock-based subdivision of the pre-Cryogenian time scale appears in the online version of this article, at https://doi.org/10.1144/jgs2020 -222; see also the original green paper of the article, at https://eartharxiv.org /repository/view/1712/.

2
Saharan Glaciers

1. Deynoux, M., and Trompette, R., "Late Precambrian mixtites: Glacial and/or non-glacial? Dealing especially with the mixtites of West Africa," discussion, *American Journal of Science* 276 (1976): 1302–1315.

2. Thomson, J., "On the geology of the island of Islay," *Transactions of the Geological Society of Glasgow* 5 (1877): 220–222. The reports of ancient glacial strata were first made in 1871 in a talk by James Thomson, F.G.S., "On

the stratified rocks of Islay," which is recorded in a report of the 41st Meeting of the British Association for the Advancement of Science, Edinburgh (John Murray, 1872), 110–111.

3. Heezen, B. C., and Ewing, M., "Turbidity currents and submarine slumps, and the 1929 Grand Banks Earthquake," *American Journal of Science* 250 (1952): 849–873.

4. Harland, W. B., "Evidence of Late Precambrian glaciation and its significance," in *Problems in Palaeoclimatology*, ed. A. E. M. Nairn (Interscience, 1964), 119–149.

5. Schermerhorn, L. J. G., "Late Precambrian mixtites: Glacial and/or non-glacial," *American Journal of Science* 274 (1974): 673–824.

6. Reusch, H., "Skuringmærker og morængrus eftervist i Finnmarken fra en periode meget ældre end 'istiden'" [Glacial striae and boulder-clay in Norwegian Lapponie from a period much older than the last ice age], *Norges Geologiske Undersøkelse* [Geological Survey of Norway] 1 (1891): 78–85. These disputes are outlined in a paper by Laajoki, K., "New evidence of glacial abrasion of the late Proterozoic unconformity around Varangerfjorden, northern Norway," in *Precambrian Sedimentary Environments: Modern Approach to Ancient Depositional Systems*, ed. W. Alterman and P. Corcoran, International Association of Sedimentologists, Special Publication 33 (2002): 405–436.

7. The periglacial aeolian deposits and supersurfaces of Mali are fully described in Deynoux, M., Kocurek, G., Proust, J. N., "Late Proterozoic periglacial aeolian deposits on the West African Platform, Taoudeni Basin," *Sedimentology* 36 (1989): 531–549.

3

Meltwater Plume

1. The Ediacaran Period was ratified in 2004 by the International Union of Geological Sciences (IUGS), as outlined in Knoll, A. H., Walter, M. R., Narbonne, G. M., Christie-Blick, N., "A new period for the geologic time scale," *Science* 305 (2004): 621–622.

2. Shields, G. A., Deynoux, M., Strauss, H., Paquet, H., Nahon, D., "Barite-bearing cap dolostones of the Taoudeni Basin, northwest Africa: Sedimentary and isotopic evidence for methane seepage after a Neoproterozoic glaciation," *Precambrian Research* 153 (2007): 209–235.

3. James, N. P., Narbonne, G. M., Kyser, T. K., "Late Neoproterozoic cap carbonates: Mackenzie Mountains, Northwestern Canada: Precipitation and global glacial meltdown," *Canadian Journal of Earth Sciences* 38 (2001): 1229–1262.

4. Crockford, P. W., Wing, B. A., Paytan, A., Hodgskiss, M. S. W., May-field, K. K., Hayles, J. A., Middleton, J. E., Ahm, A.-S. C., Johnston, D. T., Caxito, F., Uhlein, G., Halverson, G. P., Eickmann, B., Torres, M., Horner, T. J., "Barium-isotopic constraints on the origin of post-Marinoan barites," *Earth and Planetary Science Letters* 519 (2019): 234–244.

5. The film *The Day After Tomorrow* has inspired a number of critiques from scientists who despair at Hollywood's interpretation of how natural catastrophes are likely to unfold. In one paper, Sybren Drijfhout used a climate model to predict what would really happen if ever the Atlantic Meridional Overturning Circulation (AMOC) stopped: Drijfhout, S., "Competition between global warming and an abrupt collapse of the AMOC in Earth's energy imbalance," *Scientific Reports* 5 (2015): 14877, https://doi.org/10.1038/srep14877.

6. The "plume world" was first outlined in Shields, G. A., "Neoproterozoic cap carbonates: A critical appraisal of existing models and the plume-world hypothesis," *Terra Nova* 17 (2005): 299–310. The notion that global oceans remained stratified for tens of thousands of years during and after Cryogenian deglaciation has since been supported by a string of recent papers, including Liu, C., Wang, Z., Raub, T. D., Macdonald, F. A., Evans, D. A., "Neoproterozoic cap-dolostone deposition in stratified glacial meltwater plume," *Earth and Planetary Science Letters* 404 (2014): 22–32; Yang, J., Jansen, M. F., Macdonald, F. A., Abbot, D. S., "Persistence of a freshwater surface ocean after a snowball Earth," *Geology* 45 (2017): 615–618; Yu, W., Algeo, T. J., Zhou, Q., Du, Y., Wang, P., "Cryogenian cap carbonate models: A review and critical assessment," *Palaeogeography, Palaeoclimatology, Palaeoecology* 552 (2020): 109727.

7. Shields, G. A., Deynoux, M., Culver, S. J., Brasier, M. D., Affaton, M. D., Affaton, P., Vandamme, D., "Neoproterozoic glaciomarine and cap dolostone facies of the southwestern Taoudeni Basin (Walidiala Valley, Senegal/Guinea, NW Africa)," *Comptes Rendus Geoscience* 339 (2007): 186–199.

4

Frozen Greenhouse

1. Fairchild, I. J., "Balmy shores and icy wastes: The paradox of carbonates associated with glacial deposits in Neoproterozoic times," in *Sedimentology Review/1*, ed. V. Paul Wright (1993): 1–16.

2. Spencer, A. M., "Late Pre-Cambrian glaciation in Scotland," *Geological Society of London Memoirs* 6 (1971).

3. Fairchild, I. J., Spencer, A. M., Ali, D. O., Anderson, R. P., Boomer, I.,

Dove, D., Evans, J. D., Hambrey, M. J., Howe, J., Sawaki, Y., Shields, G. A., Skelton, A., Tucker, M. E., Wang, Z., Zhou, Y., "Tonian-Cryogenian boundary sections of Argyll, Scotland," *Precambrian Research* 319 (2018): 37–64.

4. Key pre-1990 paleomagnetism papers include Harland, W. B., and Bidgood, D. E. T., "Palaeomagnetism in some Norwegian sparagmites and the late pre-Cambrian ice age," *Nature* 184 (1959): 1860–1862; and Embleton, B. J. J., and Williams, G. E., "Low latitude of deposition for late Precambrian periglacial varvites in South Australia: Implications for palaeoclimatology," *Earth and Planetary Science Letters* 79 (1986): 419–430. The fold test that confirmed a low latitude for glaciation was reported in an abstract in 1987: Sumner, D. Y., Kirschvink, J. L., Runnegar, B., "Soft-sediment palaeomagnetic field tests of late Precambrian glaciogenic sediments," abstract, *Eos (Transactions of the American Geophysical Union)* 68 (1987): 1251.

5. The term "Snowball Earth" was first used by Joe Kirschvink in a short book chapter: Kirschvink, J. L., "Late Proterozoic low-latitude glaciation: The snowball Earth," in *The Proterozoic Biosphere*, ed. J. W. Schopf and C. Klein (Cambridge: Cambridge University Press, 1992), 51–52. The central tenet of the hypothesis (high atmospheric carbon dioxide levels, but global ice cover) has been widely supported by a number of geochemical studies, starting with Bao, H., Lyons, J. R., Zhou, C., "Triple oxygen isotope evidence for elevated CO_2 levels after a Neoproterozoic glaciation," *Nature* 453 (2008): 504–506; Bao, H., Fairchild, I. J., Wynn, P. M., Spötl, C., "Stretching the envelope of past surface environments: Neoproterozoic glacial lakes from Svalbard," *Science* 323 (2009): 119–122.

6. Kirschvink's Snowball Earth hypothesis was reinvigorated by a series of papers stemming from one in particular: Hoffman, P. F., Kaufman, A. J., Halverson, G. P., Schrag, D. P., "A Neoproterozoic Snowball Earth," *Science* 281 (1998): 1342–1346.

7. Jacques-Joseph Ebelmen's discovery and Urey's rediscovery of the workings of the long-term global carbon cycle: Ebelmen, J. J., "Sur les produits de la décomposition des espèces minérales de la famille des silicates," *Annales des Mines* 3 (1845): 3–66; Urey, H. C., "On the early chemical history of the Earth and the origin of life," *Proceedings of the National Academy of Sciences* 38 (1952): 351–363.

8. For a nice historical overview of the discovery of the global carbon cycle, see Galvez, M. E., and Gaillardet, J., "Historical constraints on the origins of the carbon cycle concept," *Comptes Rendus Geoscience* 344 (2012): 549–567.

9. Kasemann, S. A., Hawkesworth, C. J., Prave, A. R., Fallick, A. E., Pearson, P. N., "Boron and calcium isotope composition in Neoproterozoic carbonate rocks from Namibia: Evidence for extreme environmental change," *Earth and Planetary Science Letters* 231 (2005): 73–86.

10. The connection between seafloor aragonite fans and iron inhibition was first described by Dawn Sumner and John Grotzinger in Sumner, D. Y., and Grotzinger, J. P., "Were kinetics of Archean calcium carbonate precipitation related to oxygen concentration?" *Geology* 24 (1996): 119–122.

5
Clocks in Rocks

1. The connections between weathering, glaciation, and environmental change began to be elucidated in a series of papers, such as Derry, L. A., Kaufman, A. J., Jacobsen, S. B., "Sedimentary cycling and environmental change in the late Proterozoic: Evidence from stable and radiogenic isotopes," *Geochimica et Cosmochimica Acta* 56 (1992): 1317–1329; Kaufman, A. J., Knoll, A. H., Jacobsen, S. B., "The Vendian record of Sr and C isotopic variations in seawater: Implications for tectonics and paleoclimate," *Earth and Planetary Science Letters* 120 (1993): 409–430.

2. The post-Sturtian rise in weathering and its effects on the biosphere were first outlined in Shields, G. A., Stille, P., Brasier, M. D., Atudorei, N. V., "Stratified oceans and oxygenation of the late Precambrian environment: A post-glacial geochemical record from the Neoproterozoic of W Mongolia," *Terra Nova* 5–6 (1997): 218–222.

3. A marked increase in chemical weathering after deglaciation is now widely supported; for example, see Wei, G., Wei, W., Wang, D., Li, T., Yang, X., Shields, G. A., Zhang, F., Lie, G., Chen, T., Yang, T., Ling, H., "Enhanced chemical weathering triggered an expansion of euxinia after the Sturtian glaciation," *Earth and Planetary Science Letters* 539, 116244 (2020).

4. Multiple step sequential leaching for strontium isotopes was introduced by Chao Liu and co-authors: Liu, C., Wang, Z., Raub, T. D., "Geochemical constraints on the origin of Marinoan cap dolostones from Nuccaleena Formation, South Australia," *Chemical Geology* 351 (2013): 95–104.

5. Recent updates on the ages and synchroneity of Cryogenian ice ages can be found in Shields et al., "A template for an improved rock-based subdivision of the pre-Cryogenian timescale," *Journal of the Geological Society* 179 (2021), https://doi.org/10.1144/jgs2020-222.

6. Initial pH estimates were made by Simone Kasemann and colleagues: Kasemann, S. A., Hawkesworth, C. J., Prave, A. R., Fallick, A. E., Pearson, P. N., "Boron and calcium isotope composition in Neoproterozoic carbonate rocks from Namibia: Evidence for extreme environmental change," *Earth and Planetary Science Letters* 231 (2005): 73–86. The findings were subsequently confirmed in a later paper, also by Simone Kasemann's research group:

Ohnemueller, F., Prave, A. R., Fallick, A. E., Kasemann, S. A., "Ocean acidification in the aftermath of the Marinoan glaciation," *Geology* 42 (2014): 1103–1106.

7. A major prediction of the Snowball Earth hypothesis was the buildup of carbon dioxide levels during and immediately following glaciation. This was first confirmed in two key papers by Huiming Bao: Bao, H., Lyons, J. R., Zhou, C., "Triple oxygen isotope evidence for elevated CO_2 levels after a Neoproterozoic glaciation," *Nature* 453 (2008): 504–506; Bao, H., Fairchild, I. J., Wynn, P. M., Spötl, C., "Stretching the envelope of past surface environments: Neoproterozoic glacial lakes from Svalbard," *Science* 323 (2009): 119–122.

8. Two key papers, since confirmed numerous times, sealed the case for a synchronous end to the Cryogenian glaciations 635 million years ago: Condon, D., Zhu, M., Bowring, S., Wang, W., Yang, A., Jin, Y., "U-Pb ages from the Neoproterozoic Doushantuo Formation, China," *Science* 308 (2004): 95–98; Hoffmann, K.-H., Condon, D. J., Bowring, S. A., Crowley, J. L., "U-Pb zircon date from the Neoproterozoic Ghaub Formation, Namibia: Constraints on Marinoan glaciation," *Geology* 32 (2004): 817–820.

9. Current estimates for the onset and end of Sturtian and Marinoan glaciations (717, 660, 650, and 635 million years ago, respectively) are appraised in Halverson, G. P., Porter, S., Shields, G. A., "The Tonian and Cryogenian periods," in *Geologic Time Scale, 2020,* ed. F. M. Gradstein, J. G. Ogg, M. D. Schmitz, G. M. Ogg (Elsevier, 2020), 495–519.

10. Age constraints on the Gaskiers glaciation and overlying fossiliferous strata are summarized in Matthews, J. J., Liu, A. G., Yang, C., McIlroy, D., Levell, B., Condon, D. J., "A chronostratigraphic framework for the rise of the Ediacaran Macrobiota: New constraints from Mistaken Point Ecological Reserve, Newfoundland," *GSA Bulletin* 133 (2021): 612–624. Nikolay Chumakov's long-standing prediction of an Ediacaran-Cambrian glaciation is presented in Chumakov, N. M., "The Baykonurian Glaciohorizon of the Late Vendian," *Stratigraphy and Geological Correlation* 17 (2009): 373–381.

6

Broken Thermostat

1. In its rather archaic language and pre-decimal currency (when sixpence was worth half a shilling, and there were twenty shillings to a pound), the quotation from Charles Dickens's book *David Copperfield* describes a universal problem of tiny annual deficits inexorably leading to bankruptcy. In Victorian England, even tiny debts could result in a person being sent to

the workhouse, or forced to emigrate. In the novel, Micawber chose to be sent to Australia, where he eventually made his fortune.

2. Hutton, J., "Theory of the Earth: Or an investigation of the laws observable in the composition, dissolution, and restoration of land upon the globe," *Transactions of the Royal Society of Edinburgh* 1, part 2 (1788): 209–304. The David Hume quotation is from his book *Dialogues Concerning Natural Religion*, which was published in London in 1779.

3. Alfred Russel Wallace, in a letter to Charles Darwin written in February 1858, compared the spontaneous outcomes of natural selection to the centrifugal governor. The only known version of the text is published in the *Journal of the Proceedings of the Linnean Society (Zoology)* 3 (1858): 53–62, titled "On the tendency of varieties to depart indefinitely from the original type," which was read to the society on July 1, 1858.

4. Joseph Priestley introduced carbonic acid in his classic paper, Priestley, J., "Directions for impregnating water with fixed air in order to communicate to it the peculiar spirit and virtues of Pyrmont water, and other mineral waters of a similar nature," which was printed in London in 1772 for J. Johnson (22 pages).

5. James Lovelock's Gaia theory was first brought to wider attention in his classic book *Gaia: A New Look at Life on Earth*, published by Oxford University Press in 1979. The quotation is from page 209 of James Hutton's book *Theory of the Earth: Or an Investigation of the Laws Observable in the Composition, Dissolution, and Restoration of Land upon the Globe*, which was published in 1788 by the Royal Society of Edinburgh.

6. "If we are alone in the universe, then it is an awful waste of space": This marvelous tongue-in-cheek line is from Carl Sagan's science fiction novel *Contact*, published by Simon and Schuster in 1985.

7. The Faint Young Sun Paradox was first described in Sagan, C., and Mullen, G., "Earth and Mars: Evolution of atmospheres and surface temperatures," *Science* 177 (1972): 52–56; the Gaia hypothesis was formulated in Lovelock, J. E., "Gaia as seen through the atmosphere," *Atmospheric Environment* 6 (1972): 579–580, and developed with Lynn Margulis: Lovelock, J. E., and Margulis, L., "Atmospheric homeostasis by and for the biosphere: The Gaia hypothesis," *Tellus Series A* 26 (1974): 2–10.

8. The first complete description of the long-term global carbon cycle was in Ebelmen, J. J., "Sur les produits de la décomposition des espèces minérales de la famille des silicates," *Annales des Mines* 3 (1845): 3–66; followed by an attempt at quantification: Ebelmen, J. J., "Sur la décomposition des roches," *Annales des Mines* 4ᵉ série (1847): 627–654.

9. The evolution of lichen and carbon concentration mechanisms may have helped sequester carbon dioxide during the Tonian Period, as discussed

further by, among others, Lenton, T. M., and Watson, A. J., "Biotic enhancement of weathering, atmospheric oxygen and carbon dioxide in the Neoproterozoic," *Geophysical Research Letters* 31 (2004): L05202.

10. The coral reef effect was first proposed, and its application to the Cryogenian glaciation outlined, by Wolfgang Berger in Berger, W. H., "Increase of carbon dioxide in the atmosphere during deglaciation: The coral reef hypothesis," *Naturwissenschaften* 69 (1982): 87–88; and applied to the Cryogenian glaciations by Ridgwell, A. J., Kennedy, M. J., Caldeira, K., "Carbonate deposition, climate stability, and Neoproterozoic ice ages," *Science* 302 (2003): 859–862.

11. Tectonic quiescence during the Cryogenian Period has been proposed by McKenzie, N. R., Hughes, N. C., Gill, B. C., Myrow, P. M., "Plate tectonic influences on Neoproterozoic–early Paleozoic climate and animal evolution," *Geology* 42 (2014): 127–130; and by Mills, B. J. W., Scotese, C. R., Walding, N. G., Shields, G. A., Lenton, T. M., "Elevated CO_2 degassing rates prevented the return of Snowball Earth during the Phanerozoic," *Nature Communications* 8 (2017): 1–7.

12. Organic ballast as a means of sequestering carbon dioxide during the Tonian Period has been proposed many times since the classic paper by Knoll, A. H., Hayes, J. M., Kaufman, A. J., Swett, K., Lambert, I. B., "Secular variation in carbon isotope ratios from upper Proterozoic successions for Svalbard and East Greenland," *Nature* 321 (1986): 832–838.

13. The so-called "fire and ice" basalt weathering hypothesis was first framed by Goddéris, Y., Donnadieu, Y., Nédélec, A., Dupré, B., Dessert, C., Gerard, A., Ramstein, G., François, L. M., "The Sturtian 'snowball' glaciation: Fire and ice," *Earth and Planetary Science Letters* 211 (2003): 1–12. It was subsequently modified to take into account the dual role of basalt weathering on both carbonate and organic burial: Horton, F., "Did phosphorus derived from the weathering of large igneous provinces fertilize the Neoproterozoic ocean?" *Geochemistry, Geophysics, Geosystems* 16 (2015): 1723–1738; and Gernon, T. M., Hincks, T. K., Tyrrell, T., Rohling, E. J., Palmer, M. R., "Snowball Earth ocean chemistry driven by extensive ridge volcanism during Rodinia breakup," *Nature Geoscience* 9 (2016): 242–250.

14. The Great Oxidation Event is marked by the disappearance of detrital pyrite from the marine sedimentary record, indicating that atmospheric oxygen levels have been sufficient since at least 2.2 billion years ago to oxidize iron sulfide during subaerial weathering and subsequent transport. The GOE is tightly associated with the work of Heinrich "Dick" Holland. For example, see Holland, H. D., "Volcanic gases, black smokers, and the Great Oxidation Event," *Geochimica et Cosmochimica Acta* 66 (2002): 3811–3826. For an up-to-date summary of this event, see Poulton, S. W., Bekker, A., Cum-

ming, V. M., Zerkle, A. L., Canfield, D. E., Johnston, D. T., "A 200-million-year delay in permanent atmospheric oxygenation," *Nature* 592 (2021): 232–236.

15. The significance of the relatively muted variability of the mid-Proterozoic carbon isotope record was noted early on by Brasier, M. D., and Lindsay, J. F., "A billion years of environmental stability and the emergence of eukaryotes: New data from northern Australia," *Geology* 26 (1998): 555–558.

16. The origin of the term "boring billion" to describe the relative quiescence of the earth's Middle Age is described in Martin Brasier's book: Brasier, M. D., *Secret Chambers: The Inside Story of Cells and Complex Life* (Oxford University Press, 2012).

17. A negative feedback between atmospheric oxygenation and organic carbon weathering likely resulted in carbon isotope stability during much of the Proterozoic Eon as deduced in an astute numerical modeling paper by Daines, S. J., Mills, B. J. W., Lenton, T. M., "Atmospheric oxygen regulation at low Proterozoic levels by incomplete oxidative weathering of sedimentary organic carbon," *Nature Communications* 8 (2017): 14379. This suggests that carbon isotope excursions, which bracket and occasionally interrupt the "boring billion" interval of relative stability, were produced via imbalances within the sulfur cycle.

18. The notion that an enormous organic carbon capacitor persisted in the surface environment throughout much of the Neoproterozoic Era was first raised in a mathematical modeling study by Rothman, D. H., Hayes, J. M., Summons, R. E., "Dynamics of the Neoproterozoic carbon cycle," *Proceedings of the National Academy of Sciences* 100 (2003): 8124–8129; the smoking guns for such a capacitor are the aforementioned negative carbon isotope excursions: Shields, G. A., "Carbon and carbon isotope mass balance in the Neoproterozoic Earth system," *Emerging Topics in Life Science* 2 (2018): 257–265.

7
Fossil Records

1. Charles Darwin wrote *On the Origin of Species, by Means of Natural Selection* in 1859 (published by John Murray). In the first edition, Darwin referred to the Silurian System, whereas today we use the term Cambrian. Martin Brasier discusses the different approaches to Darwin's Dilemma of Lyell, Sollas, and Daly in his book *Darwin's Lost World* (Oxford University Press, 2009).

2. Shore et al. make the case for a specific taxonomic affinity for some late Ediacaran animals: Shore, A. J., Wood, R. A., Butler, I. B., Zhuravlev, A. Yu., McMahon, S., Curtis, A., Bowyer, F. T., "Ediacaran metazoan reveals lophotrochozoan affinity and deepens root of Cambrian explosion," *Science Advances* 7 (2021): eabf2933. A more general animalian (eumetazoan) affinity for Ediacaran fossils can be traced back to at least 574 million years ago: Dunn, F. S., Liu, A. G., Grazhdankin, D. V., Vixseboxse, P., Flannery-Sutherland, J., Green, E., Harris, S., Wilby, P. R., Donoghue, P. C. J., "The developmental biology of *Charnia* and the eumetazoan affinity of the Ediacaran rangeomorphs," *Science Advances* 7 (2021): eabe0291.

3. Although molecular phylogeny trees place the last common ancestor of all extant bilaterian animals in the early Ediacaran or even Cryogenian (dos Reis, M., Thawornattana, Y., Angelis, K., Telford, M. J., Donoghue, P. C. J., "Uncertainties in the timing of origin of animals and the limits of precision in molecular timescales," *Current Biology* 25 [2015]: 2939–2950), the oldest fossil evidence is no older than about 555 million years, comprising slug-like *Kimberella* fossils in Australia and Russia (Fedonkin, M. A., and Waggoner, B. M., "The Late Precambrian fossil *Kimberella* is a mollusc-like bilaterian organism," *Nature* 388 [1997]: 868–871) and likely bilaterian traces in Australia and China (Evans, S. D., Hughes, I. V., Gehling, J. G., Droser, M. L., "Discovery of the oldest bilaterian from the Ediacaran of South Australia," *Proceedings of the National Academy of Sciences* 117 [2020]: 7845–7850; Chen, Z., Zhou, C., Yuan, X., Xiao, S., "Death march of a segmented and trilobate bilaterian elucidates early animal evolution," *Nature* 573 [2019]: 412–415).

4. Bill Schopf first reported putative bacteria from the Archean in Schopf, J. W., and Packer, B. M., "Early Archean (3.3-billion to 3.5-billion-year-old) microfossils from Warrawoona Group, Australia," *Science* 237 (1987): 7–73; while the rebuttal from Martin Brasier was published as Brasier, M. D., Green, O. R., Jephcoat, A. P., Kleppe, A. K., Van Kranendonk, M. J., Lindsay, J. F., Steele, A., Grassineau, N. V., "Questioning the evidence for Earth's oldest fossils," *Nature* 416 (2002): 76–81.

5. Organic macrofossils have been reported from 2.0–1.6-billion-year-old rocks around the world, for example in the form of coils (*Grypania*), discs, or even fronds: Han, T. M., and Runnegar, B., "Megascopic eukaryotic algae from the 2.1-million-year-old Negaunee iron-formation, Michigan," *Science* 257 (1992): 232–235; Zhu, S., and Chen, H., "Megascopic multicellular organisms from the 1700-million-year-old Tuanshanzi Formation in the Jixian area, North China," *Science* 270 (1995): 620–622.

6. The most convincing large eukaryotes are from China and have been dated at about 1.56 billion years ago: Zhu, S., Zhu, M., Knoll, A. H., Yin, Z.,

Zhao, F., Sun, S., Qu, Y., Shi, M., Liu, H., "Multicellular eukaryotes from the 1.56-million-year-old Gaoyuzhuang Formation in North China," *Nature Communications* 7 (2016): 11500.

7. The earliest ornamental acritarchs (likely eukaryote) microfossils are from rocks aged about 1.65 billion years: Miao, L., Moczydlowska, M., Zhu, S., Zhu, M., "New record of organic-walled, morphologically distinct microfossils from the late Paleoproterozoic Changcheng Group in the Yanshan Range, North China," *Precambrian Research* 321 (2019): 172–198.

8. Evidence for eukaryotic diversification before and during the Cryogenian Period comes from molecular biomarker studies: Brocks, J. J., Jarrett, A. J. M., Sirantoine, E., Hallmann, C., Hoshino, Y., Liyanage, T., "The rise of algae in Cryogenian oceans and the emergence of animals," *Nature* 548 (2017): e1700887.

9. Red algae are the oldest eukaryotic taxa that can readily be identified in the fossil record, at about 1.05 billion years ago; see Butterfield, N. J., "Bangiomorpha pubescens n. gen., n. sp.: Implications for the evolution of sex, multicellularity, and the Mesoproterozoic/Neoproterozoic radiation of eukaryotes," *Paleobiology* 26 (2000): 386–404. Red algae are followed by chlorophyte green algae; see Tang, Q., Pang, K., Yuan, X., Xiao, S., "A one-billion-year-old multicellular chlorophyte," *Nature Ecology and Evolution* 4 (2020): 543–549.

10. Susannah Porter and Andy Knoll reported the oldest convincing heterotrophic eukaryotes in the form of testate amoebae in Porter, S. M., and Knoll, A. H., "Testate amoebae in the Neoproterozoic Era: Evidence from vase-shaped microfossils in the Chuar Group, Grand Canyon," *Paleobiology* 26 (2000): 360–385. For the first scales and the rise of eukaryvorous predators, see Cohen, P. A., Schopf, J. W., Butterfield, N. J., Kudryavtsev, Macdonald, F. A., "Phosphate biomineralization in mid-Neoproterozoic protists," *Geology* 39 (2011): 539–542; and Porter, S. M., "The rise of predators," *Geology* 39 (2011): 607–608.

11. The idea that choanocytes were ancestral to all animals comes from Dujardin, F., *Histoire naturelle des Zoophytes, Infusoires, comprenant la physiologie et la classification de ces animaux et la manière de les étudier à l'aide du microscope* (Librairie encyclopédique de Roret, 1841).

12. The sponge biomarker debate is covered in a recent paper, Bobrovskiy, I., Hope, J. M., Nettersheim, B. J., Volkman, J. K., Hallmann, C., Brocks, J. J., "Algal origin of sponge sterane biomarkers negates the oldest evidence for animals in the rock record," *Nature Ecology and Evolution* 5 (2021): 165–168. Body fossil evidence for animals is entirely absent for the Cryogenian Period, while even possible fossil protists, namely ciliates, have recently been

reinterpreted as algae: Cohen, P. A., Vizcaino, M., Anderson, R. P., "Oldest fossil ciliates from the Cryogenian glacial interlude reinterpreted as possible red algal spores," *Palaeontology* 63, no. 6 (2020): 941–950.

13. The Cryogenian Period has been identified as the crucible of animal evolution in a number of molecular phylogeny studies that incorporate a sophisticated and cautious approach to the quantification of uncertainties: dos Reis, M., Thawornattana, Y., Angelis, K., Telford, M. J., Donoghue, P. C. J., Yang, Z., "Uncertainty in the timing of origin of animals and the limits of precision in molecular timescales," *Current Biology* 25 (2015): 2939–2950.

14. On altruism: Boyle, R. A., Lenton, T. M., Williams, H. T. P., "Neoproterozoic 'snowball Earth' glaciations and the evolution of altruism," *Geobiology* 5 (2007): 337–349.

15. The "Ediacaran" discoid fossil *Aspidella* has been identified in the nonglacial interval of the Cryogenian Period in northwest Canada: Burzynski, G., Dececchi, T. A., Narbonne, G. M., Dalrymple, R. W., "Cryogenian *Aspidella* from northwestern Canada," *Precambrian Research* 336 (2020): 105507.

16. Large spiny acritarchs are characteristic fossils of the early Ediacaran and were identified as the likely resting cysts of animals in Yin, L., Zhu, M., Knoll, A. H., Yuan, X., Zhang, J., Hu, J., "Doushantuo embryos preserved inside diapause egg cysts," *Nature* 446 (2007): 661–663; and Cohen, P. A., Knoll, A. H., Kodner, R. B., "Large spinose microfossils in Ediacaran rocks as resting stages of early animals," *Proceedings of the National Academy of Sciences* 106 (2009): 6519–6524.

17. *Aspidella* was discovered by Alexander Murray in 1868 as reported in Murray, A., *Report upon the Geological Survey of Newfoundland for 1868* (St. John's, Newfoundland, Canada: Robert Winton, 1869); see page 11. The famous paleontologist Charles Doolittle Walcott poured cold water on the discovery in a paper titled "Pre-Cambrian fossiliferous formations" that was published in the *Bulletin of the Geological Society of America* in 1899. On page 231, Walcott quotes "Mr. G.F. Matthew, of Saint John, New Brunswick": "I have seen *Aspidella terranovica* in the museum at Ottawa and doubt its organic origin. It seems to me a slickensided mud concretion striated by pressure."

18. Roger Mason's discovery of *Charnia* was first reported in Ford, T. D., "Precambrian fossils from Charnwood Forest, Leicestershire," *Proceedings of the Yorkshire Geological Society* 3 (1958): 211–217; see also Kenchington, C. G., Harris, S. J., Vixseboxse, P. B., Pickup, C., Wilby, P. R., "The Ediacaran fossils of Charnwood Forest: Shining new light on a major biological revolution," *Proceedings of the Geologists' Association* 129 (2018): 264–277.

19. Matthews, J. J., Liu, A. G., Yang, C., McIlroy, D., Levell, B., Condon, D. J., "A chronostratigraphic framework for the rise of the Ediacaran Macro-

biota: New constraints from Mistaken Point Ecological Reserve, Newfoundland," *GSA Bulletin* 133 (2021): 612–624. The Shuram anomaly is dated vari-
ously between about 570 and 551 million years ago but is likely to be from
no longer than 10 million years ago, so needs better constraints; see, for example: Rooney, A. D., Cantine, M. D., Bergmann, K. D., Gomez-Perez, I., Al
Baloushi, B., Boag, T. H., Busch, J. F., Sperling, E. A., Strauss, J. V., "Calibrat-
ing the coevolution of Ediacaran life and environment," *Proceedings of the
National Academy of Sciences* 117 (2020): 16824–16830.

20. Martin Brasier's discovery of *Haootia* is reported in Liu, A. G., Matthews, J. J., Menon, L. R., McIlroy, D., Brasier, M. D., "*Haootia quadriformis*
n. gen, n. sp., interpreted as a muscular cnidarian impression from the Late
Ediacaran Period (approx. 560 Ma)," *Proceedings of the Royal Society B* 281
(2014): 20141202. An even more convincing example of a probable cnidarian
was subsequently found in England, and of about the same age. It is described in Dunn, F. S., Kenchington, C. G., Parry, L. A., Clark, J. W., Kendall,
R. S., Wilby, P. R., "A crown-group cnidarian from the Ediacaran of Charnwood Forest, UK," *Nature Ecology & Evolution* 6 (2022): 1095–1104, https://
doi.org/10.1038/s41559-022-01807-x.

21. The exquisite 3D forms of Newfoundland Ediacarans have been described by various research groups, most notably that of Guy Narbonne: Narbonne, G. M., "The Ediacara biota: Neoproterozoic origin of animals and
their ecological systems," *Annual Review of Earth and Planetary Sciences* 33
(2005): 421–442.

22. An osmotrophic metabolism was first inferred in Laflamme, M.,
Xiao, S., Kowalezski, M., "Osmotrophy in modular Ediacara organisms," *Proceedings of the National Academy of Sciences* 106 (2009): 14438–14443; and
supported mathematically in Hoyal Cuthill, J. F., Conway Morris, S., "Fractal
branching organizations of Ediacaran rangeomorph fronds reveal a lost
Proterozoic body plan," *Proceedings of the National Academy of Sciences* 111
(2014): 13122–13126.

8
Oxygen Rise

1. Oparin, A. I., *Proiskhozhdenie zhizni* (Moscow: Moskovskii Rabochii,
1924), expanded and published in book form in Russian in 1936, translated
and published in English as *The Origin of Life* in 1938; Haldane, J. B. S., "The
origin of life," *The Rationalists Annual* 148 (1929): 3–10; Miller, S. L., "A production of amino acids under possible primitive conditions," *Science* 117
(1953): 528–529. These developments in Darwin's primordial soup have been

challenged, not least because it is hard to envisage how the requisite thermo-dynamic gradients that lubricate metabolisms could have evolved under such homogeneous conditions, as explained by Nick Lane in his book *The Vital Question: Energy, Evolution, and the Origins of Complex Life* (New York: W.W. Norton, 2015).

2. Nursall, J. R., "Oxygen as a prerequisite to the origin of the meta-zoan," *Nature* 183 (1959): 1170–1172.

3. Hutton, J., "Theory of the Earth: Or an investigation of the laws ob-servable in the composition, dissolution, and restoration of land upon the globe," *Transactions of the Royal Society of Edinburgh* 1, part 2 (1788): 209–304. John Playfair recalled his thoughts on angular unconformities in his reading of a "Biographical Account of the Late Dr James Hutton, F.R.S. Edin." (read 1803), *Transactions of the Royal Society of Edinburgh* 5 (1805): 71–73: "We made for a high rocky point or head-land, the SICCAR. . . . On landing at this point, we found that we actually trode [*sic*] on the primeval rock. . . . It is here a micaceous schistus, in beds nearly vertical, highly indurated, and stretching from S. E. to N. W. The surface of this rock . . . has thin covering of red horizontal sandstone laid over it. . . . Here, therefore, the immediate contact of the two rocks is not only visible, but is curiously dissected and laid open by the action of the waves. . . . On us who saw these phenomena for the first time, the impression will not easily be forgotten. The palpable evidence presented to us, of one of the most extraordinary and important facts in the natural history of the earth, gave a reality and substance to those theoretical speculations, which, however probable had never till now been directly au-thenticated by the testimony of the senses. . . . What clearer evidence could we have had of the different formation of these rocks, and of the long inter-val which separated their formation, had we actually seen them emerging from the bosom of the deep? . . . The mind seemed to grow giddy by looking so far into the abyss of time; and while we listened with earnestness and admiration to the philosopher who was now unfolding to us the order and series of these wonderful events, we became sensible how much farther rea-son may sometimes go than imagination can venture to follow."

4. The stoichiometric effect of pyrite burial on oxygen budgets was elu-cidated by early pioneers of a numerical treatment of the long-term carbon cycle; for example, Garrels, R. M., and Lerman, A., "Phanerozoic cycles of sedimentary carbon and sulfur," *Proceedings of the National Academy of Sci-ences* 78 (1981): 4652–4656.

5. Kaufman, A. J., Knoll, A. H., Jacobsen, S. B., "The Vendian record of Sr and C isotopic variations in seawater: Implications for tectonics and pa-leoclimate," *Earth and Planetary Science Letters* 120 (1993): 409–430.

6. The relevance of Ce depletion for high rates of chemical weathering

and surface oxygenation immediately following Sturtian glaciation was first pointed out in Shields, "Global environmental change following the late Precambrian ice ages: A rare earth element approach," abstract, International Geological Congress, Beijing, 1996; Shields, G. A., Stille, P., Brasier, M. D., Atudorei, N. V., "Stratified oceans and oxygenation of the late Precambrian environment: A post-glacial geochemical record from the Neoproterozoic of W Mongolia," *Terra Nova* 5–6 (1997): 218–222. See also discussion of these findings in Nick Lane's book *Oxygen: The Molecule That Made the World* (Oxford University Press, 2002).

7. The Neoproterozoic Oxygenation Event (NOE) was proposed in Shields, G. A., and Och, L., "The case for a Neoproterozoic oxygenation event: Geochemical evidence and biological consequences," *GSA Today* 21 (2011): 4–11; Och, L., and Shields, G. A., "The Neoproterozoic event: Environmental perturbations and biogeochemical cycling," *Earth Science Reviews* 110 (2012): 26–57.

8. Redox-sensitive transition metal abundances record oxygenation pulses during the Ediacaran Period: Scott, C., Lyons, T. W., Bekker, A., Shen, Y., Poulton, S. W., Chu, X., Anbar, A., "Tracing the stepwise oxygenation of the Proterozoic ocean," *Nature* 452 (2008): 456–459; Sahoo, S. K., Planavsky, N. J., Kendall, B., Wang, X., Shi, X., Scott, C., Anbar, A. D., Lyons, T. W., Jiang, G., "Ocean oxygenation in the wake of the Marinoan glaciation," *Nature* 489 (2012): 546–549.

9. The iron partitioning approach to determining whether water columns were locally anoxic or oxic, widely referred to as iron speciation, was established by Raiswell, R., and Canfield, D. E., "Sources of iron for pyrite formation in marine sediments," *American Journal of Science* 298 (1998): 219–245; and refined by Poulton, S. W., and Canfield, D. E., "Development of a sequential extraction procedure for iron: Implications for iron partitioning in continentally derived particulates," *Chemical Geology* 214 (2005): 209–221.

10. Don Canfield proposed that anoxia was the normal condition of the sub-surface oceans for most of the Proterozoic Eon in Canfield, D. E., "A new model for Proterozoic ocean chemistry," *Nature* 396 (1998): 450–453; and provided evidence for oxygenation of the deeper marine environment during the mid-Ediacaran in Canfield, D. E., Poulton, S. W., Narbonne, G. M., "Late-Neoproterozoic deep-ocean oxygenation and the rise of animal life," *Science* 315 (2007): 92–95.

11. The eight-armed *Eoandromeda* was reported in Zhu, M., Gehling, J. G., Xiao, S., Zhao, Y., Droser, M. L., "Eight-armed Ediacara fossil preserved in contrasting taphonomic windows from China and Australia," *Geology* 36 (2008): 867–870. Molybdenum isotope evidence for ocean oxygenation through this interval comes from Chen, X., Ling, H., Vance, D., Shields, G. A.,

Zhu, M., Poulton, S. W., Och, L., Jiang, S., Li, D., Cremonese, L., Archer, C., "Rise to modern levels of ocean oxygenation coincided with the Cambrian radiation of animals," *Nature Communications* 6 (2015): 1–7.

12. The "evolutionary innovation then oxygenation" model is exemplified in papers by, for example, Butterfield, N. J., "Oxygen, animals and oceanic ventilation: An alternative view," *Geobiology* 7 (2009): 1–7; Sperling, E. A., Pisani, D., Peterson, K. J., "Poriferan paraphyly and its implications for Precambrian palaeobiology," in *The Rise and Fall of the Ediacaran Biota*, ed. P. Vickers-Rich and P. Komarower, Geological Society, London, Special Publications 286 (2007): 355–368; Lenton, T. M., Boyle, R. A., Poulton, S. W., Shields, G. A., Butterfield, N. J., "Co-evolution of eukaryotes and ocean oxygenation in the Neoproterozoic Era," *Nature Geoscience* 7 (2014): 257–264.

13. Shields, G. A., and Zhu, M., "Biogeochemical changes across the Ediacaran-Cambrian transition in South China," *Precambrian Research* 225 (2013): 1–6; Boyle, R. A., Dahl, T. W., Dale, A. W., Shields, G. A., Zhu, M., Brasier, M. D., Lenton, T. M., "Stabilization of the coupled oxygen and phosphorus cycles by the evolution of bioturbation," *Nature Geoscience* 7 (2014): 671–676.

14. The oldest metazoan reefs were reported in Penny, A. M., Wood, R. A., Curtis, A., Bowyer, F., Tostevin, R., Hoffmann, K.-H., "Ediacaran metazoan reefs from the Nama Group, Namibia," *Science* 344 (2014): 1504–1506.

15. A return to anoxia following the oxygenation of circa 560 million years ago (Shuram anomaly interval) is shown by the uranium isotope record: Zhang, F., Xiao, S., Kendall, B., Romaniello, S. J., Cui, H., Meyer, M., Gilleaudeau, G. J., Kaufman, A. J., Anbar, A. D., "Extensive marine anoxia during the terminal Ediacaran Period," *Science Advances* 4, eaan8983 (2018); Tostevin, R., Clarkson, M. O., Gangl, S., Shields, G. A., Wood, R. A., Bowyer, F., Penny, A. M., Stirling, C. H., "Uranium isotope evidence for an expansion of anoxia in terminal Ediacaran oceans," *Earth and Planetary Science Letters* 506 (2019): 104–112.

16. Oscillating redox conditions are now thought to be characteristic of the Ediacaran-Cambrian transition interval and may have contributed to the Cambrian explosion by driving evolutionary radiation and bottleneck events: He, T., Zhu, M., Mills, B. J. W., Wynn, P. M., Zhuravlev, A. Yu., Tostevin, R., Pogge von Strandmann, P. A. E., Yang, A., Poulton, S. W., Shields, G. A., "Possible links between extreme oxygen perturbations and the Cambrian radiation of animals," *Nature Geoscience* 12 (2019): 468–472; Wei, G., Planavsky, N. J., Tarhan, L. G., He, T., Wang, D., Shields, G. A., Wei, W., Ling, H., "Highly dynamic marine redox state through the Cambrian explosion highlighted by authigenic $\delta^{234}U$ records," *Earth and Planetary Science Letters* 544, 116361 (2020).

17. The phosphorus cycling ratio is described in Lenton, T. M., Dutreuil, S., Latour, B., "Life on Earth is hard to spot," *The Anthropocene Review* 7 (2020): 248–272.

9
Limiting Nutrients

1. Khabarov, N., and Obersteiner, M., "Global phosphorus fertilizer market and national policies: A case study revisiting the 2008 price peak," *Frontiers in Nutrition* 4 (2017), http://doi.org/10.3389/fnut.2017.00022.

2. The theory or law of minimum was first formulated by Carl Sprengel, for example in Sprengel, C., *Meine Erfahrungen im Gebiete der allgemeinen und speciellen Pflanzen-Cultur*, 3 vols. (Baumgärtners Buchhandlung: Leipzig, 1847–1852).

3. Xiao, S., Zhang, Y., Knoll, A. H., "Three-dimensional preservation of algae and animal embryos in a Neoproterozoic phosphorite," *Nature* 391 (1998): 553–558.

4. The taxonomic affinity of the Weng'an embryos has evolved through time with interpretations ranging from animalian to bacterial: Bailey, J. V., Joye, S. B., Kalanetra, K. M., Flood, B. E., Corsetti, F. A., "Evidence of giant sulphur bacteria in Neoproterozoic phosphorites," *Nature* 445 (2007): 198–201. Embryos were first reported within common acritarch casings in the paper by Yin, L., Zhu, M., Knoll, A. H., Yuan, X., Zhang, J., Hu, J., "Doushantuo embryos preserved inside diapause egg cysts," *Nature* 446 (2007): 661–663.

5. A holozoan (protistan) affinity is gaining traction for the interpretation of many of the embryos. See, for example, Huldtgren, T., Cunningham, J. A., Yin, C., Stampanoni, M., Marone, F., Donoghue, P. C. J., Bengtson, S., "Fossilized nuclei and germination structures identify Ediacaran 'animal embryos' as encysting protists," *Science* 334 (2011): 1696–1699; Yin, Z., Sun, W., Liu, P., Zhu, M., Donoghue, P., "Developmental biology of *Helicoforamina* reveals holozoan affinity, cryptic diversity, and adaptation to heterogeneous environments in the early Ediacaran Weng'an biota (Doushantuo formation, South China)," *Science Advances* 6 (2020): eabb0083. The possibility that sponges were among the Weng'an embryos has been mooted several times, most recently in the paper Yin, Z., Sun, W., Reitner, J., Zhu, M., "New holozoans with cellular resolution from the early Ediacaran Weng'an biota, Southwest China," *Journal of the Geological Society* 179 (2022), http://doi.org/10.1144/jgs2021-061.

6. Subcellular organelles and nuclei have been convincingly demon-

strated by Zongjun Yin: Yin, Z., Cunningham, J. A., Vargas, K., Bengtson, S., Zhu, M., Donoghue, P. C. J., "Nuclei and nucleoli in embryo-like fossils from the Ediacaran Weng'an biota," *Precambrian Research* 301 (2017): 145–151; Sun, W., Yin, Z., Cunningham, J. A., Liu, P., Zhu, M., Donoghue, P. C. J., "Nucleus preservation in early Ediacaran Weng'an embryo-like fossils, experimental taphonomy of nuclei and implications for reading the eukaryote fossil record," *Interface Focus* 10 (2020), http://doi.org/10.1098/rsfs.2020.0015.

7. Proximity of the Weng'an embryos to a water column redox boundary was inferred in a sulfur isotope and rare earth element study: Shields, G. A., Kimura, H., Yang, J., Gammon, P., "Sulphur isotopic evolution of Neoproterozoic-Cambrian seawater: New francolite-bound sulphate $\delta^{34}S$ data and a critical appraisal of the existing record," *Chemical Geology* 204 (2004): 163–182.

8. Oxygenation of the seafloor may have led to increased phosphorus burial (deoxygenation) due to two separate mechanisms: Higgins, J. A., Fischer, W. W., Schrag, D. P., "Oxygenation of the oceans and sediments: Consequences for the seafloor carbonate factory," *Earth and Planetary Science Letters* 284 (2009): 25–33; and Shields, G. A., and Zhu, M., "Biogeochemical changes across the Ediacaran-Cambrian transition in South China," *Precambrian Research* 225 (2013): 1–6; Boyle, R. A., Dahl, T. W., Dale, A. W., Shields, G. A., Zhu, M., Brasier, M. D., Lenton, T. M., "Stabilization of the coupled oxygen and phosphorus cycles by the evolution of bioturbation," *Nature Geoscience* 7 (2014): 671–676. However, increased sulfate levels may lead to more efficient phosphorus recycling and deposition of phosphorite giant deposits: Laakso, T. A., Sperling, E. A., Johnston, D. T., Knoll, A. H., "Ediacaran reorganization of the marine phosphorus cycle," *Proceedings of the National Academy of Sciences* 117 (2020): 11961–11967.

9. The order of appearance of biominerals and their relationship to contemporaneous ocean composition are outlined in Zhuravlev, A. Yu, and Wood, R. A., "Eve of biomineralization: Controls on skeletal mineralogy," *Geology* 36 (2008): 923–926; Porter, S. M., "Calcite and aragonite seas and the de novo acquisition of carbonate skeletons," *Geobiology* 8 (2010): 256–277.

10. Walcott's Lipalian interval: Walcott, C. D., *Cambrian Geology and Paleontology II*, Smithsonian Miscellaneous Collections (Smithsonian Institution, 1914), c. 57, 14.

11. High erosion rates around the Ediacaran-Cambrian transition are evidenced in the Sr isotope record: Shields, G. A., "A normalised seawater strontium isotope curve: Possible implications for Neoproterozoic-Cambrian weathering rates and the further oxygenation of the Earth," *eEarth* 2 (2007): 35–42; and the tectonic/zircon abundance records: Zhu, Z., Campbell, I. H.,

Allen, C. A., Burnham, A. D., "S-type granites: Their origin and distribution through time as determined from detrital zircons," *Earth and Planetary Science Letters* 536, 116140 (2020).

12. The correspondence between zircon abundance and the supercontinent cycle became clear through the work of Kent Condie over two decades: Condie, K. C., and Puetz, S. J., "Time series analysis of mantle cycles Part II: The geologic record in zircons, large igneous provinces and mantle lithosphere," *Geoscience Frontiers* 10 (2019): 1327–1336.

13. For an overview of zircon isotope records, see Spencer, C. J., "Continuous continental growth as constrained by the sedimentary record," *American Journal of Science* 320 (2020): 373–401.

14. Squire, R., Campbell, I., Allen, C., Wilson, C., "Did the Transgondwanan Supermountain trigger the explosive radiation of animals on Earth?" *Earth and Planetary Science Letters* 250 (2006): 116–133; preceded by Brasier, M. D., and Lindsay, J. F., "Did supercontinental amalgamation trigger the 'Cambrian Explosion'?" in *The Ecology of the Cambrian Radiation* (Columbia University Press, 2000), 68–89.

15. The oldest fossil evidence for oceangoing nitrogen fixation is preserved in Chinese rocks of about 1,000 million years ago: Pang, K., Tang, Q., Chen, L., Wan, B., Niu, C., Yuan, X., Xiao, S., "Nitrogen-fixing heterocystous cyanobacteria in the Tonian Period," *Current Biology* 28 (2018): 616–626. Such a late occurrence was predicted based on phylogenomic considerations: Sanchez-Baracaldo, P., Ridgwell, A., Raven, J. A., "A Neoproterozoic transition in the marine nitrogen cycle," *Current Biology* 24 (2014): 652–657.

16. Nitrogen isotope evidence for a fluctuating but enlarged nitrate pool during the Ediacaran and Cambrian Periods comes from Ader, M., Sansjofre, P., Halverson, G. P., Busigny, V., Trinidade, R. I. F., Kunzmann, M., Noguiera, A. C. R., "Ocean redox structure across the Late Neoproterozoic Oxygenation Event: A nitrogen isotope perspective," *Earth and Planetary Science Letters* 396 (2014): 1–13; Wang, X., Jiang, G., Shi, X., Peng, Y., Morales, D., "Nitrogen isotope constraints on the early Ediacaran ocean redox structure," *Geochimica et Cosmochimica Acta* 240 (2018): 220–235; Wang, D., Ling, H., Struck, U., Zhu, X., He, T., Yang, B., Gamper, A., Shields, G. A., "Coupling of ocean redox and animal evolution during the Ediacaran-Cambrian transition," *Nature Communications* 9 (2019): 1–8.

17. Martin Brasier commonly remarked on the role of unusually early phosphogenesis in fossil preservation, especially in Proterozoic rocks, for example in Brasier, M. D., "Phosphogenic events and skeletal preservation across the Precambrian-Cambrian boundary interval," *Phosphorite Research and Development*, ed. A. J. G. Northolt and I. Jarvis, Geological Society, Lon-

don, Special Publications 52 (1990): 289–303; Battison, L., and Brasier, M. D., "Remarkably preserved prokaryote and eukaryote microfossils within lake phosphates of the Torridonian Group, NW Scotland," *Precambrian Research* 196–197 (2012): 204–217.

10

A Pinch of Salt

1. The Messinian Salinity Crisis came to people's notice through the paper Hsu, K. J., Ryan, W. B. F., Cita, M. B., "Late Miocene desiccation of the Mediterranean," *Nature* 242 (1973): 240–244. Ken Hsu popularized the idea in his book *The Mediterranean Was a Desert: A Voyage of the* Glomar Challenger (Princeton University Press, 1987). The Caspian Sea Model for the upper gypsum is explained in more detail in Andreeto, F., Aloisi, G., Raad, F., et al., "Freshening of the Mediterranean Salt Giant: Controversies and certainties around the terminal (Upper Gypsum and Lago Mare) phases of the Messinian Salinity Crisis," *Earth Science Reviews* 216, 103577 (2021).

2. Only a few compilations of evaporite giants through time have been made, much of which is based on Russian authors M. A. Zharkov and A. B. Ronov. See, for example, Warren, J. K., "Evaporites through time: Tectonic, climatic and eustatic controls in marine and non-marine deposits," *Earth Science Reviews* 98 (2010): 217–268.

3. Hsu, K. J., Oberhänsli, H., Gao, J. Y., Shu, S., Chen, H., Krähenbühl, U., "'Strangelove ocean' before the Cambrian explosion," *Nature* 316 (1985): 809–811.

4. The "Strangelove ocean" scenario of export production collapse after mass extinctions, evidenced by ^{12}C depletion, had a long-lasting effect on interpretations of negative $\delta^{13}C$ excursions: Kump, L. R., "Interpreting carbon-isotope excursions: Strangelove oceans," *Geology* 19 (1991): 299–302; Hoffman, P. F., Kaufman, A. J., Halverson, G. P., Schrag, D. P., "A Neoproterozoic Snowball Earth," *Science* 281 (1998): 1342–1346.

5. After doing the rounds with high-impact journals like *Nature,* the first report of the Shuram anomaly finally came out in Burns, S. J., and Matter, A., "Carbon isotopic record of the latest Proterozoic from Oman," *Eclogae Geologicae Helvetiae* 86 (1993): 595–607.

6. Shuram-equivalent excursions have been reported from all around the world, including Australia, Russia, and Scotland, but the greatest number of sections from different settings has been reported from China, for example: Lu, M., Zhu, M., Zhang, J., Shields, G. A., Li, G., Zhao, F., Zhao, M.,

"The DOUNCE event at the top of the Ediacaran Doushantuo Formation, South China: Broad stratigraphic occurrence and non-diagenetic origin," *Precambrian Research* 225 (2013): 86–109.

7. The list of negative carbon isotope anomalies has grown long, with the most convincing examples being the pre-Marinoan Trezona of ~645 million years ago (Rose, C. V., Swanson-Hysell, N. L., Husson, J. M., Poppick, L. N., Cottle, J. M., Schoene, B., Maloof, A. C., "Constraints on the origin and relative timing of the Trezona δ^{13}C anomaly below the end-Cryogenian glaciation," *Earth and Planetary Science Letters* 319–320 [2012]: 241–250); the pre-Sturtian Islay and Garvellach of ~735 and ~720 million years ago, respectively (Fairchild, I. J., Spencer, A. M., Ali, D. O., Anderson, R. P., Boomer, I., Dove, D., Evans, J. D., Hambrey, M. J., Howe, J., Sawaki, Y., Shields, G. A., Skelton, A., Tucker, M. E., Wang, Z., Zhou, Y., "Tonian-Cryogenian boundary sections of Argyll, Scotland," *Precambrian Research* 319 [2018]: 37–64); the Bitter Springs of ~800 million years ago (Halverson, G. P., Maloof, A. C., Schrag, D. P., Dudas, F. O., Hurtgen, M., "Stratigraphy and geochemistry of a ca 800 Ma negative carbon isotope interval in northeastern Svalbard," *Chemical Geology* 237 [2007]: 5–27); the Majiatun of ~930 million years ago (Park, H., Zhai, M., Yang, J., Peng, P., Kim, J., Zhang, Y., Kim, M., Park, U., Feng, L., "Deposition age of the Sangwon Supergroup in the Pyongnam basin [Korea] and the Early Tonian negative carbon isotope interval," *Yanshi Xuebao* 32 [2016]: 2181–2195); the Gaoyuzhuang of ~1.56 billion years ago (Zhang, K., Zhu, X., Wood, R. A., Shi, Y., Gao, Z., Poulton, S. W., "Oxygenation of the Mesoproterozoic ocean and the evolution of complex eukaryotes," *Nature Geoscience* 11 [2018]: 1110–1120); and potentially also earlier anomalies, for example ~1.9 billion years ago (Kump, L. R., Junium, C., Arthur, M. A., Brasier, A., Fallick, A., Melezhik, V., Lepland, A., Crune, A. E., Luo, G., "Isotopic evidence for massive oxidation of organic matter following the Great Oxidation Event," *Science* 334 [2011]: 1694–1696).

8. The first paper to point out that negative carbon isotope excursions could not be sustained even under modern levels of high oxygenation was Bristow, T. F., and Kennedy, M. J., "Carbon isotope excursions and the oxidant budget of the Ediacaran atmosphere and ocean," *Geology* 36 (2008): 863–866.

9. Various studies have tried to address the Shuram anomaly quandary, for example, Bjerrum, C. J., and Canfield, D. E., "Towards a quantitative understanding of the late Neoproterozoic carbon cycle," *Proceedings of the National Academy of Sciences* 108 (2011): 5542–5547.

10. Wortman, U. G., and Paytan, A., "Rapid variability of seawater chemistry over the past 130 million years," *Science* 337 (2012): 334–336.

11. Shields, G. A., Mills, B. J. W., Zhu, M., Raub, T. D., Daines, S., Lenton, T. M., "Unique Neoproterozoic carbon isotope excursions sustained by coupled evaporite dissolution and pyrite burial," *Nature Geoscience* 12 (2019): 823–827.

12. The silicate weathering feedback was first mooted in Walker, J. C. G., Hays, P. B., Kasting, J. F., "A negative feedback mechanism for the long-term stabilization of Earth's surface temperature," *Journal of Geophysical Research* 86 (1981): 9776–9782, whereby "Carbon dioxide partial pressure and Earth's greenhouse effect are . . . buffered by the temperature dependence of the rate of carbon dioxide consumption in the weathering of silicate minerals." The GEOCARBSULF model originated in the early 1980s as BLAG, named for its authors Bob Berner, Anthony Lasaga, and Bob Garrels. BLAG was the first global model to quantify all conceivable processes that govern CO_2 fluxes over long time scales. Mathematical simplifications and the addition of isotopic data turned BLAG into GEOCARB in 1991, for example: Berner, R. A., "GEOCARBSULF: A combined model for Phanerozoic atmospheric O_2 and CO_2," *Geochimica et Cosmochimica Acta* 70 (2006): 5653–5664. The origins and evolution of such models are reviewed in Mills, B. J. W., Krause, A. J., Scotese, C. R., Hill, D. J., Shields, G. A., Lenton, T. M., "Modelling the long-term carbon cycle, atmospheric CO_2 and Earth surface temperature from late Neoproterozoic to present day," *Gondwana Research* 67 (2019): 172–186.

13. The COPSE model turns the GEOCARBSULF approach on its head in the sense that isotopic trends do not drive the model's results but instead are outputs used to test the model's validity. This avoids circular reasoning and the inevitable conclusion that negative excursions are impossible. The COPSE model was developed by Noah Bergman, Tim Lenton, and Andy Watson in 2004 (Bergman, N. M., Lenton, T. M., Watson, A. J., "COPSE: A new model of biogeochemical cycling over Phanerozoic time," *American Journal of Science*, 304 [2004]: 397–437), and is continually being developed, for example: Lenton, T. M., Daines, S. J., Mills, B. J. W., "COPSE reloaded: an improved model of biogeochemical cycling over Phanerozoic time," *Earth Science Reviews* 178 (2018): 1–28.

14. Rising sulfate and other oxyanion concentrations in seawater through the Shuram anomaly have been reported in many studies, for example: Fike, D. A., Grotzinger, J. P., Pratt, L. M., Summon, R. E., "Oxidation of the Ediacaran ocean," *Nature* 444 (2006): 744–747; Kendall, B., Komiya, T., Lyons, T. W., Bates, S. M., Gordon, G. W., Romaniello, S. J., Jiang, G., Creaser, R. A., Xiao, S., McFadden, K., Sawaki, Y., Tahata, M., Shu, D., Han, J., Li, Y., Chu, X., Anbar, A.D., "Uranium and molybdenum isotope evidence for an episode of widespread ocean oxygenation during the late Ediacaran period," *Geochimica et Cosmochimica Acta* 156 (2015): 173–193.

15. Calcium toxicity resulting from a rise in ocean calcium concentration has long been speculated to have been a cause of the Cambrian explosion, for example: Simkiss, K., "Biomineralization and detoxification," *Calcified Tissue Research* 24 (1977): 199–200. More recently, the choice of biomineral by specific taxa has been linked to the ocean's composition at the time of their evolution: Zhuravlev, A. Yu., and Wood, R. A., "Eve of biomineralization: Controls on skeletal mineralogy," *Geology* 36 (2008): 923–926. High Ca concentrations in the late Ediacaran ocean have been linked to orogeny: Peters, S. E., and Gaines, R. R., "Formation of the 'Great Unconformity' as a trigger for the Cambrian explosion," *Nature* 484 (2012): 363–366, although evaporite weathering provides the specific mechanism for increasing Ca, while maintaining calcium carbonate saturation in the oceans: Shields, G. A., Mills, B. J. W., Zhu, M., Raub, T. D., Daines, S., Lenton, T. M., "Unique Neoproterozoic carbon isotope excursions sustained by coupled evaporite dissolution and pyrite burial," *Nature Geoscience* 12 (2019): 823–827.

16. The redox reversal from boom to bust after the Shuram anomaly is best seen in Tostevin, R., Clarkson, M. O., Gangl, S., Shields, G. A., Wood, R. A., Bowyer, F., Penny, A. M., Stirling, C. H., "Uranium isotope evidence for an expansion of anoxia in terminal Ediacaran oceans," *Earth and Planetary Science Letters* 506 (2019): 104–112. Interestingly, the fossil record also shows that the latest Ediacaran times saw animals relegated to the more oxygenated shallow marine environments: Tostevin, R., Wood, R. A., Shields, G. A., Poulton, S. W., Guilbaud, R., Bowyer, F., Penny, A., He, T., Curtis, A., Hoffmann, K.-H., Clarkson, M. O., "Low-oxygen waters limited habitable space for early animals," *Nature Communications* 7 (2016): 1–9; Xiao, S., Chen, Z., Zhou, C., Yuan, X., "Surfing in and on microbial mats: Oxygen-related behavior of a terminal Ediacaran bilaterian animal," *Geology* 47 (2019): 1054–1058.

17. Evolutionary radiations and bottlenecks caused by fluctuating oxygen levels are reported in He, T., Zhu, M., Mills, B. J. W., Wynn, P. M., Zhuravlev, A. Yu., Tostevin, R., Pogge von Strandmann, P. A. E., Yang, A., Poulton, S. W., Shields, G. A., "Possible links between extreme oxygen perturbations and the Cambrian radiation of animals," *Nature Geoscience* 12 (2019): 468–472.

11

Biosphere Resilience

1. Darwin's life in London is told in great style in Adrian Desmond and James R. Moore's biography *Darwin*, published by Penguin in 1992.

2. Classic nineteenth-century papers mentioned here include Phillips, J., *Life on Earth: Its Origin and Succession* (London: Macmillan, 1860); Logan,

W. E., "On the division of Azoic rocks of Canada into Huronian and Lauren-tian," *Proceedings of the American Association for the Advancement of Science* (1857): 44–47; Sedgwick, A., "On the older Palæozoic (Protozoic) rocks of North Wales," *Quarterly Journal of the Geological Society* 1 (1845): 5–22.

3. The K-Pg meteorite impact was first evidenced in Alvarez, L. W., Al-varez, W., Asaro, F., Michel, H. V., "Extraterrestrial cause for the Cretaceous-Tertiary extinction," *Science* 208 (1980): 1095–1108. However, most mass ex-tinctions seem to be related to periods of sustained greenhouse forcing by volcanism and ocean anoxia: Courtillot, V., *Evolutionary Catastrophes: The Science of Mass Extinction* (Cambridge University Press, 1999); Hallam, A., and Wignall, P. B., *Mass Extinctions and Their Aftermath* (Oxford University Press, 2000).

4. The size of the Messinian evaporite event was recently updated using seismic visualization to about a million cubic kilometers, in Haq, B., Gorini, C., Baur, J., Moneron, J., Rubino, J.-L., "Deep Mediterranean's Mes-sinian evaporite giant: How much salt?" *Global and Planetary Change* 184 (2020): 103052.

5. Jacobs, E., Weissert, H., Shields, G. A., Stille, P., "The Monterey event in the Mediterranean: A record from shelf sediments of Malta," *Paleoceanog-raphy* 11 (1996): 717–728.

6. The world's descent into glaciation through the Cenozoic Era is re-corded in the oxygen isotope compositions of foraminifera, for example the classic "Zachos curve": Zachos, J., Pagani, M., Sloan, L., Thomas, E., Billips, K., "Trends, rhythms, and aberrations in global climate 65 Ma to present," *Science* 292 (2001): 686–693; and its updated version: Westerhold, T., Marwan, N., Drury, A. J., Liebrand, et al., "An astronomically dated record of Earth's cli-mate and predictability over the last 66 million years," *Science* 369 (2020): 1383–1387.

7. The conventional silicate weathering paradigm of climate regulation has been challenged from several viewpoints that highlight the significant effects of orogeny on the global carbon cycle, for example in papers on sili-cate weathering: Raymo, M. E., and Ruddiman, W. F., "Tectonic forcing of late Cenozoic climate," *Nature* 359 (1992): 117–122; pyrite weathering: Torres, M. A., West, A. J., Li, G., "Sulphide oxidation and carbonate dissolution as a source of CO_2 over geological timescales," *Nature* 507 (2014): 346–349; and evaporite weathering: Shields, G. A., and Mills, B. J. W., "Evaporite weather-ing and deposition as a long-term climate forcing mechanism," *Geology* 49 (2021): 299–303.

8. Benton, M. J., *When Life Nearly Died: The Greatest Mass Extinction of All Time* (Thames & Hudson, 2003).

9. Benton, M. J., "Hyperthermal-driven mass extinctions: Killing models

during the Permian-Triassic mass extinction," *Philosophical Transactions of the Royal Society A* 376 (2018): 20170076.

10. Kump, L. R., "Prolonged Late Permian–Early Triassic hypothermal: Failure of climate regulation?" *Philosophical Transactions of the Royal Society A* 376 (2018): 20170078.

11. He, T., Dal Corso, J., Newton, R. J., Wignall, P. B., Mills, B. J. W., Todaro, S., Di Stefano, P., Turner, E. C., Jamieson, R. A., Randazzo, V., Rigo, M., Jones, R. E., Dunhill, A. M., "An enormous sulfur isotope excursion indicates marine anoxia during the end-Triassic mass extinction," *Science Advances* 6 (2020): eabb6704.

12. The history of the Deep Carbon Observatory is outlined in a book by Robert Hazen, *Symphony in C: Carbon and the Evolution of (Almost) Everything* (W.W. Norton, 2019).

13. The emergence of continents by the end of the Archean is a popular notion, for example: Dhuime, B., Wuestefeld, B., Hawkesworth, C. J., "Emergence of modern continental crust about 3 billion years ago," *Nature Geoscience* 8 (2015): 552–555.

14. Blättler, C. L., Claire, M. W., Prave, A. R., Kirsimae, K., Higgins, J. A., Medvedev, P. V., Romashkin, A. E., Rychanchik, D. V., Zerkle, A. L., Paiste, K., Kreitsmann, T., Millar, I. L., Hayles, J. A., Bao, H., Turchyn, A. V., Warke, M. R., Lepland, A., "Two-billion-year-old evaporites capture Earth's great oxidation," *Science* 360 (2018): 320–323.

15. Further evidence for astonishingly high, Phanerozoic-like sulfate concentrations in Paleoproterozoic seawater comes from all around the world, for example: Schröder, S., Bekker, A., Beukes, N. J., Strauss, H., van Niekerk, H. S., "Rise in seawater sulphate concentration associated with the Paleoproterozoic positive carbon isotope excursion: Evidence from sulphate evaporites in the ~2.2–2.1 Gyr shallow-marine Lucknow Formation, South Africa," *Terra Nova* 20 (2008): 108–117.

16. The effects of supercontinent amalgamation and denudation on carbon isotopes and oxygenation are described in more detail in Shields, G. A., and Mills, B. J. W., "Tectonic controls on the long-term carbon isotope mass balance," *Proceedings of the National Academy of Sciences* 114 (2017): 4318–4323. The role of sulfur in oxidation events is described for the first time in this book; see also Shields, G. A., "The role of sulphate in Earth's great oxidation events," abstract, Goldschmidt 2021 Virtual Conference, no. 8111, July 4–9, 2021.

17. The natural fission reactor in Gabon was discovered in the early 1970s; see Bodu, R., Bouzigues, H., Morin, N., Pfiffelmann, J. P., "Sur l'existence d'anomalies isotopiques rencontrées dans l'uranium du Gabon," *Comptes Rendus de l'Académie des Sciences, Paris* 275 (1972): 1731–1734.

18. The claim for aerobic organisms, preserved in mineral pyrite, was first made in a paper in *Nature* by El Albani, A., Bengtson, S., Canfield, D. E., Bekker, A., et al., "Large colonial organisms with coordinated growth in oxygenated environments 2.1 Gyr ago," *Nature* 466 (2010): 100–104. The evidence for an aerobic environment comes from the low ratio in rock samples between highly reactive and total iron, but this seems unlikely considering the sheer amount of pyrite in the samples. Motility is claimed in the form of trace fossils in a later paper: El Albani, A., Mangano, M. G., Buatois, L. A., Bengtson, S., et al., "Organism motility in an oxygenated shallow-marine environment 2.1 billion years ago," *Proceedings of the National Academy of Sciences* 116 (2019): 3431–3436. Sulfur isotope values confirm the surprisingly high sulfate concentrations in seawater around this time, but also suggest euxinia, inconsistent with an aerobic environment.

19. Replaying the "tape of life" is a central theme in *Wonderful Life: The Burgess Shale and the Nature of History*, by Stephen Jay Gould (W.W. Norton, 1989). Toby Tyrrell's modeling study of the earth's habitability is Tyrrell, T., "Chance played a role in determining whether Earth stayed habitable," *Communications Earth and Environment* 1 (2020): 61.

12

Time's Arrow

1. Tim Lenton's interview with James Lovelock, "James Lovelock Centenary: The Future of Global Systems Thinking," *Global Systems Institute, University of Exeter*, July 29–31, 2019, https://www.lovelockcentenary.info. In August 2022, James Lovelock finally ceased to be an active participant in Gaia at the grand age of 103.

2. Don Anderson published his views of how our planet works in his comprehensive book, Anderson, D. L., *New Theory of the Earth* (New York: Cambridge University Press, 2007).

3. Evolution through symbiosis is explored in Nick Lane's book *Power, Sex, Suicide: Mitochondria and the Meaning of Life* (Oxford University Press, 2005).

4. A great synthesis of how Gaia theory evolved is given in Tim Lenton and Andy Watson's book *Revolutions That Made the Earth* (Oxford University Press, 2011).

5. Jones, S., *Almost Like a Whale: The Origin of Species Updated* (BCA, 1999).

6. Tyrrell, T., *On Gaia: A Critical Investigation of the Relationship Between Life and Earth* (Princeton University Press, 2013); Waltham, D., *Lucky*

Planet: Why Earth is Exceptional—and What That Means for Life in the Universe (Icon Books Limited, 2014).

7. "Life is nothing but an electron looking for a place to rest" is from the biochemist Albert Szent-Györgyi, in "Bioelectronics," *Science* 161 (1968): 988–990.

8. Delayed oxygenation was proposed in Dahl, T. W., Hammarlund, E. U., Anbar, A. D., Bond, D. P. G., Gill, B. C., Gordon, G. W., Knoll, A. H., Nielsen, A. T., Schovsbo, N. H., Canfield, D. E., "Devonian rise in atmospheric oxygen correlated to the radiations of terrestrial plants and large predatory fish," *Proceedings of the National Academy of Sciences* 107 (2010): 17911–17915.

9. Underlying reasons for continuing surface oxygenation are explored in Hayes, J. M., and Waldbauer, J. R., "The carbon cycle and associated redox processes through time," *Philosophical Transactions of the Royal Society B* 361 (2006), http://doi.org/10.1098/rstb.2006.1840; Mills, B. J. W., Lenton, T. M., Watson, A. J., "Proterozoic oxygen rise linked to shifting balance between seafloor and terrestrial weathering," *Proceedings of the National Academy of Sciences* 111 (2014): 9073–9078.

10. The estimate of 10–100 times is from Lenton, T. M., Dutreuil, S., Latour, B., "Life on Earth is hard to spot," *The Anthropocene Review* 7 (2020): 248–272. The paper by Stern, R. J., and Gerya, T., mentioned here is "Earth evolution, emergence, and uniformitarianism," *GSA Today* 31 (2021): 32–33. The term "Biogeodynamics" is attributed to Aubrey Zerkle, a geochemist at the University of St. Andrews in Scotland; see Zerkle, A. L., "Biogeodynamics: Bridging the gap between surface and deep Earth processes," *Philosophical Transactions of the Royal Society A* 376 (2018): 20170401, http://doi.org/10.1098/rsta.2017.0401.

11. Lane, N., *The Vital Question: Energy, Evolution, and the Origins of Complex Life* (W.W. Norton, 2015).

12. A cool early earth is supported by numerous geochemical models, as described, for example, in Krissansen-Totton, J., Arney, G. N., Catling, D. C., "Constraining the climate and ocean pH of the early Earth with a geological carbon cycle model," *Proceedings of the National Academy of Sciences* 115 (2018): 4105–4110.

13. The Deep Sea Drilling Project (DSDP) helped identify variability in the depths at which carbonate biominerals dissolve in the oceans, for example: Berger, W. H., "Deep sea carbonates: Dissolution facies and age-depth constancy," *Nature* 236 (1972): 392–395.

14. Carbon concentration mechanisms in cyanobacteria evolved early on to shield the enzyme RUBISCO from high ambient oxygen levels following the GOE, despite relatively high CO_2: Hurley, S. J., Wing, B. A., Jasper, C. E.,

Hill, N. C., Cameron, J. C., "Carbon isotope evidence for the global physiol-ogy of Proterozoic cyanobacteria," *Science Advances* 7 (2021): eabc8998.
    15. The 0.3% figure is from Lenton, T. M., Dutreuil, S., Latour, B., "Life on Earth is hard to spot," *The Anthropocene Review* 7 (2020): 248–272.

Index